THE GSM EVOLUTION

THE GSM EVOLUTION

Mobile Packet Data Services

Peter Stuckmann

Aachen University, Germany

JOHN WILEY & SONS, LTD

Telephone (+44) 1243 779777

E-mail (for orders and customer service enquiries): cs-books@wiley.co.uk

Visit our Home Page on www.wileyeurope.com or www.wiley.com

Other Wiley Editorial Offices

John Wiley & Sons, Inc., 111 River Street, Hoboken, NJ 07030, USA

Jossey-Bass, 989 Market Street, San Francisco, CA 94103-1741, USA

Wiley-VCH Verlag GmbH, Boschstr. 12, D-69469 Weinheim, Germany

John Wiley & Sons Australia Ltd, 33 Park Road, Milton, Queensland 4064, Australia

John Wiley & Sons (Asia) Pte Ltd, 2 Clementi Loop #02-01, Jin Xing Distripark, Singapore 129809

John Wiley & Sons Canada Ltd, 22 Worcester Road, Etobicoke, Ontario, Canada M9W 1L1

Library of Congresss Cataloging-in-Publication Data

Stuckmann, Peter.
The GSM Evolution : mobile packet data services / Peter Stuckmann
p. cm.
Based on a tutorial presented at several international conferences and as training courses for companies. The book has been expanded from a chapter contributed by Stuckmann for Bernhard Walke's Mobile radio network.
Includes bibliographical references and index.
ISBN 0-470-84855-3
1. Global system for mobile communications. 2. Packet switching (Data transmission)
I. Title.

TK5103.483.S78 2002
621.382—dc21

2002031120

British Library Cataloguing in Publication Data

A catalogue record for this book is available from the British Library

ISBN 0-470-84855-3

Produced from LaTeX files supplied by the author.
Printed and bound in Great Britain by T. J. International Ltd, Padstow, Cornwall.
This book is printed on acid-free paper responsibly manufactured from sustainable forestry in which at least two trees are planted for each one used for paper production.

Contents

Preface

With the development of the *Global System for Mobile communication* (GSM) standard for digital cellular mobile radio networks in the late 1980s in Europe and the introduction of these systems in the 1990s a new mass market with several million subscribers worldwide has been created. Beside the growth of subscriber numbers, the technological evolution of GSM is also continuing. New services and applications have been developed and standardized and are presently integrated into GSM networks. While its principal use is for mobile telephony, mobile data services are becoming more and more popular. The success of text messaging in Europe and the growth in both subscriber numbers and usage of the i-mode service in Japan, which enables the delivery of Internet-like content on mobile phones, has led to very high market potential estimations for packet-oriented mobile data services. Licenses for *Third-Generation* (3G) systems based on *Universal Mobile Telecommunication System* (UMTS), which are seen as the successor of GSM-based *Second-Generation* (2G) systems, were granted in 1999–2002 in Europe to offer the needed radio capacity for data services with higher data rates and enhanced speech services.

The standards *General Packet Radio Service* (GPRS) and *Enhanced Data rates for GSM Evolution* (EDGE) are enhancements of the GSM standard to support packet-oriented data services with higher bit rates than are possible with GSM and to enable transparent access to fixed packet data networks. While GPRS will already provide data services, before UMTS is being deployed, both GPRS and EDGE have the potential to complement 3G networks. Since UMTS will initially only be available in areas with high subscriber density, data services will be supported outside the UMTS coverage area by GPRS with reduced quality of service and by EDGE with comparable quality of service. With the availability of multi-mode terminals supporting both GSM/EDGE and UMTS in the next years, GSM packet data services are predicted to play an important role for mobile communications far beyond 2010.

When I started my research on GSM packet data services in 1998 at the Chair of Communication Networks of Aachen University of Technology (RWTH) under the supervision of Prof. Walke, I had the advantage of sharing the comprehensive knowledge on GPRS with my predecessor and former colleague Dr. Goetz Brasche, who was involved in the standardization of GPRS in the 1990s. The RWTH has played a significant role in the development of the GPRS protocols. To support the ETSI specification groups in their choice of the protocols at the GPRS radio interface, a simulation tool, the GPRSim, has been developed to analyze the traffic performance of different candidate protocols that are also partially based on our research. After the basis for the GPRS standard was finished at the end of the 1990s, leading network operators in Europe collaborated with us in feasibility and dimensioning studies necessary for the economic introduction and evolution of GPRS/EDGE. In this framework I have further developed the GPRSim to an emulator comprising the implementation of the GPRS and EDGE protocols and load generators for typical GPRS/EDGE usage. This tool allows us to study the existing and evolved systems in their natural environments with the appropriate radio coverage, mobility and typical traffic volumes and to analyze our own approaches in the improvements and introduction of new features. The further development of this simulation tool to a comprehensive capacity planning tool and the achievement of the traffic engineering results was only possible with the support of on average about five Master students and student assistants whom I organized in a research group under my supervision. In particular I want to thank the following persons who have helped me to achieve this goal:

- Oliver Paul
- Hernrik Finck
- Jörg Franke
- Frank Müller
- Adolfo Bayo Ruiz
- Niels Ehlers
- Mirza Shahrukh Baig
- Christian Hoymann
- Jens Wolff
- Tim Irnich
- Bernd Scheunert
- Oliver Möller

This textbook is based on a tutorial that I have presented at several international conferences and as training courses for companies. Both the book and the training course are intended for all those who want to gain a deeper knowledge of GPRS and EDGE communication principles, protocols and applications as well as traffic issues of cellular packet radio networks. Addressed are students and researchers in related engineering disciplines and those working in industry or with network operators, who use or apply this technology. The book has been expanded from a chapter that I had authored for Prof. Walke's book "Mobile Radio Networks". I especially want to thank Prof. Walke for permission to reuse some of these passages as well as for his very active supervision and support of my research at his chair.

Additionally, I want to thank my colleagues Christian Hoymann and Tim Irnich for their support in developing the illustrations for the training course and the book.

Aachen, September 2002 *Peter Stuckmann*

1
Introduction

Since the introduction of digital cellular radio networks in the 1990s that were the successor of analog cellular telephone systems, the subscriber numbers and usage of mobile telephony services have grown exponentially worldwide. This step from first to second generation mobile radio networks has led to a number of incompatible systems for different regions.

The European standard GSM is based on the *Time Division Multiple Access* (TDMA) concept and uses eight time slots on 200 kHz carrier frequencies. GSM is operating at 900 MHz and 1800 MHz in Europe and in other countries worldwide and at 1900 MHz in America.

In the US two additional groups of systems have been put in place: the TDMA systems IS-54 (also called D-AMPS) and the further development IS-136 (Digital PCS) are in part similar to the GSM, while IS-95 (cdmaOne) is based on *Code Division Multiple Access* (CDMA). With a carrier frequency bandwidth of 1.23 MHz IS-95 systems are also called *Narrowband-CDMA* (N-CDMA) systems. IS-95 is also used in South America, Africa and Asia.

A separate standard was also developed in Japan. *Personal Digital Cellular* (PDC) is also based on TDMA and works with three time slots per 25 kHz channel at 800 MHz and 1500 MHz.

All these systems were designed to support speech services with data rates of 5-15 kbit/s. With the rapid increase of the Internet traffic volume in telecommunication networks and the success of the mobile data services *Short Message Service* (SMS) in Europe and i-mode in Japan, a growing demand for mobile data services is predicted. With this background *Third-Generation* (3G) cellular radio networks that were standardized in the 1990s will be introduced in the next few years.

1.1 The IMT-2000 Family of Systems

The *Internationl Telecommunication Union* (ITU) has defined requirements for the development of 3G systems. The main focus is the support of data services with high data rates (144 kbit/s outdoor and 2 Mbit/s indoor), efficient support of asymmetric traffic, packet-switched transmission at the air interface and high spectrum efficiency [53]. To enable *Global Multimedia Mobility* (GMM), a family of compatible systems should be defined, which results from the evolution of the existing incompatible systems. Six out of 15 proposals were accepted for the *International Mobile Telecommunications at 2000 MHz* (IMT-2000) family of systems, which can be grouped into four categories:

W-CDMA: *Wideband Code Division Multiple Access* (W-CDMA) systems comprise the *Frequency Division Duplex* (FDD) component of the *Universal Mobile Telecommunication System* (UMTS) and the US standard *cdma2000.*

TD-CDMA: The *Time Division—Code Division Multiple Access* (TD-CDMA) group of systems contains the *Time Division Duplex* (TDD) component of UMTS and the Chinese *Time Division—Synchronous Code Division Multiple Access* (TD-SCDMA) system, which is now integrated in the UMTS-TDD mode.

TDMA: The *Enhanced Data rates for GSM Evolution* (EDGE) concept, the further development of GSM and IS-136, has been chosen under the name *Universal Wireless Communications* (UWC)-136 to be accepted for the IMT-2000 family.

FD-TDMA: *Frequency Division—Code Division Multiple Access* (FD-CDMA) is the further development of the European *Digital Enhanced Cordless Telecommunications* (DECT) standard for cordless telephony and has also been accepted.

It is clear from these alternatives of IMT-2000 systems for 3G solutions that no unique standard can be expected worldwide. To realize global multimedia mobility, one important goal of IMT-2000, dual-mode or even multi-mode terminals will be needed always to be connected independently of the home-used standard of a customer. In addition, 3G terminals for a long while will need to have a 2G air interface, typically GSM/GPRS, implemented, to be able to connect independently of the roll-out status of 3G systems.

For the migration path from 2G cellular radio networks to 3G systems different scenarios are possible (see Figure 1.1). In Europe, UMTS will be introduced as the successor of GSM, while *GSM/EDGE Radio Access Networks* (GERANs) have the potential to complement *UMTS Terrestrial Radio Access Networks* (UTRANs) in areas with small subscriber density. EDGE will also probably be deployed in the *EDGE Compact* version in the US as the evolution of IS-136 systems [64].

Initially, the *European Telecommunications Standards Institute* (ETSI) has been given the responsibility for producing the UMTS and the evolved GSM standards. In 1999 the *3rd Generation Partnership Project* (3GPP) was formed to coordinate the efforts of the standardization institutes ETSI (Europe), *Association of Radio Industries and Businesses* (ARIB) (Japan), *Telecommunication Industry Association* (TIA) (USA), *Chinese Wireless Telecommunications Standards* (CWTS) (China), *Telecom-*

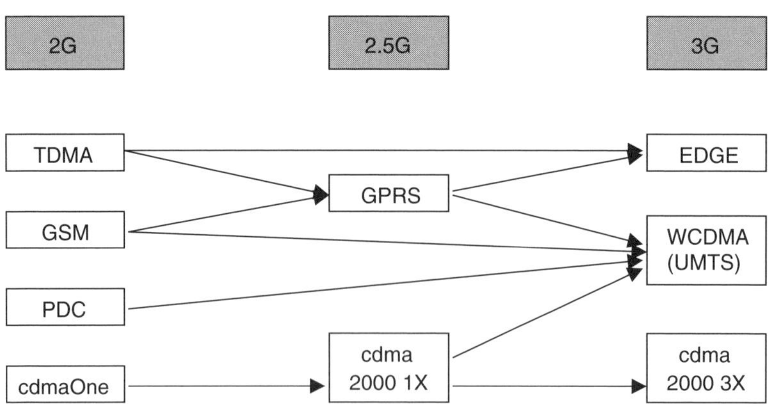

Figure 1.1: Possible migration paths from 2G to 3G systems

munication Technology Association (TTA) (Korea) and *Telecommunication Technology Committee* (TTC) (Japan).

The 3GPP is organized in several *Technical Specification Groups* (TSGs) responsible for *Core Network* (CN), *Services and System Aspects* (SSA), *Radio Access Network* (RAN), *Terminal Aspects* (TA), and *GSM/EDGE Radio Access Network* (GERAN).

For the development of cdma2000 standards the *3rd Generation Partnership Project 2* (3GPP2) was formed out of the standardization bodies ARIB, TIA, and CWTS.

1.2 GPRS and EDGE

3G mobile radio networks like UMTS will provide data services with higher data rates than are possible with 2G systems. UMTS is aiming to realize peak bit rates of up to 144 kbit/s with wide coverage, up to 384 kbit/s in hotspots, and up to 2 Mbit/s in indoor scenarios. Stepwise progress toward 3G, however, will already be performed by the extension and the further development of existing cellular systems. The advantage of such an evolutionary process is the faster availability of such services, since the infrastructure of existing 2G systems can be used. Furthermore, there is the opportunity to prepare the customers for new, so-called 3G services.

High-Speed Circuit-Switched Data (HSCSD) was introduced in some countries in 1999, and the first services based on the *General Packet Radio Service* (GPRS) have been available since 2001 in Europe.

Packet-switching means that GPRS radio resources are used only when users are actually sending or receiving data. Rather than dedicating a radio channel to a mobile data user for a fixed period of time, the available radio resource can be shared between several users. The actual number of users supported depends on the application being used and how much data is being transferred. Through multiplexing of several logical connections to one or more GSM physical channels, GPRS reaches a flexible use of channel capacity for applications with variable bit rates.

Many countries worldwide will introduce GPRS in the next few years. With these new services mobile data applications with peak bit rates of up to 117 kbit/s and typical user data rates of 25–64 kbit/s will be offered and established on the market. To realize higher data rates the ETSI has developed the EDGE standard. The packet-oriented part *Enhanced General Packet Radio Service* (EGPRS) offers a peak bit rate of up to 384 kbit/s and typical user data rates of 40–100 kbit/s by means of modified modulation, coding and medium access schemes.

1.3 Traffic Engineering

With the migration from GSM to GPRS and EGPRS, packet-switched transmission over the air interface is supported.

While for the time period following the service introduction minimal configurations were chosen supporting only a basic availability of GPRS, with increasing data traffic load in the coming years GSM/GPRS cell capacity will have to be extended. For these evolution scenarios, dimensioning guidelines are required that describe the relationship between the offered traffic and the radio resources to be allocated to reach a desired quality of service for the different applications.

For traffic engineering in circuit-switched networks the Erlang-B-Formula [82] has been successfully applied over decades, while for packet-switched cellular radio networks such an applicable traffic engineering model is still missing. The analytical description of statistical multiplexing and Internet and multimedia traffic modeling are more complex than for circuit-switched networks. Although results for packet multiplexer systems are available, e.g. for *Asynchronous Transfer Mode* (ATM) networks [63, 66, 84, 67], the specifics of the radio interface have not been included in the models so far.

Analytical and algorithmic models for the performance analysis of packet-switched radio networks are under development, but the full details of the (E)GPRS protocol stacks of the radio interface and the fixed network and of the Internet protocols, including the characteristics of TCP, currently cannot be included in the models that are analytically tractable. Since GPRS networks are presently introduced in the field, traffic engineering rules and related performance results are needed soon so that capacity and performance estimations are possible in advance of the GPRS introduction and its evolution.

Measuring the traffic performance in the existing GPRS network is not possible, since a scenario with a well-defined traffic load is hard to set up, the evaluation of the performance based on measurement is very difficult, and the analysis of the behavior of different protocol options is not possible in an existing radio network [108].

Therefore computer simulation based on the prototypical implementation (called emulation) of the GPRS protocols and the Internet protocols in combination with stochastic traffic generators for the regarded applications and models for the radio channel have been chosen as the methodology to get the needed results rapidly. These are partly shown in the last chapter of this book.

1.4 Outline

After the presentation of the evolved GSM systems with their architecture and protocols, typical existing and future applications that probably will be used over evolved GSM and 3G networks are presented. Then the traffic performance of these systems regarding the previously introduced applications is discussed.

In the second chapter I give some basics of the GSM that are necessary to understand the functionality of GPRS. The integration of the GPRS into the GSM is introduced as well.

Chapters 3 and 4 explain the architecture and the protocols of GPRS in detail. Chapter 3 focuses on the user plane. There, it is described how packets are transmitted over the air interface and transferred over the access and core network. The control plane is described in Chapter 4, which comprises the important GPRS functions for mobility management, session management and *Quality of Service* (QoS) management both in the radio access network and in the core network.

The evolution of the GSM/GPRS networks towards 3G networks is presented in Chapter 5. The modifications of the GPRS to support EDGE functionality at the air interface are presented followed by the access and core network evolution towards GERAN.

In Chapter 6 typical existing and predicted applications and the traffic characteristics together with the related service platforms are presented. After an overview of existing Internet applications like the *World Wide Web* (WWW), e-mail and *File*

Transfer Protocol (FTP), applications developed for mobile devices like *Wireless Application Protocol* (WAP)-based applications, including the *Multimedia Messaging Service* (MMS), are explained. Then future applications like Audio and Video Streaming as well as *Voice over IP* (VoIP) are described and related traffic models are proposed.

Chapter 7 introduces the simulation model that was used to gain the traffic performance results in Chapter 8. The results comprise the traffic performance of Internet applications over GPRS and EGPRS regarding fixed and on-demand channels, the traffic performance of the different evolving applications in comparison and in traffic mix scenarios, as well as results for different *Quality of Service* (QoS) management mechanisms in the radio access network and in the core network.

2 Packet Data Support in GSM Networks

2.1 GSM—Global System for Mobile Communication

This section presents the basic concept of the *Global System for Mobile communication* (GSM) to set the stage for understanding the network architecture and the radio interface that supports the General Packet Radio Service. For further information see the GSM *Technical Specifications* (GTS) or [111, 50, 88, 109].

2.1.1 GSM History

Cellular telecommunication is one of the fastest growing telecommunication applications ever developed. While the early 1980s were marked by the development of a number of national and incompatible radio networks, today cellular telecommunications represent a large and continuously increasing percentage of all new telephone subscriptions around the world.

In 1982 the Nordic *Postal, Telephone and Telegraph administration* (PTT) sent a proposal to the *Conférence Européenne des Administrations des Postes et des Télécommunications* (CEPT) to specify a common European telecommunication service at 900 MHz. A standardization group, the *Groupe Spéciale Mobile* (GSM), was established in order to formulate the specifications for this pan-European mobile cellular radio system.

There were no guidelines on how the new mobile radio system was to transmit analog or digital speech and data. The decision to develop a digital mobile radio network was not made until the development stage. But it was agreed from the beginning that the system being planned should incorporate and consider new technology from the area of telecommunications, such as *Telecommunication Standardization Sector of ITU* (ITU-T) *Common Channel Signaling System No. 7* (CCS7), *Integrated Services Digital Network* (ISDN) and the *International Standards Organization* (ISO)/*Open Systems Interconnection* (OSI) reference model. In 1986 there was a field test in Paris where the GSM group tested a number of prototypes for digital cellular radio systems. In 1987 the GSM decided on a standard that combined the best characteristics of different systems. At this time the first 17 countries signed the *Memorandum of Understanding* (MoU) committing themselves to fulfill the specifications and confirmed their commitment to introducing mobile radio based on the recommendations of the GSM. Figure 2.1 shows the milestones in the GSM history.

Later, in March 1989, the GSM working party was taken over by ETSI, and since 1991 has been called the *Special Mobile Group* (SMG). Today the abbreviation GSM stands for *Global System for Mobile Communication*, thereby underlining its claim as a worldwide standard.

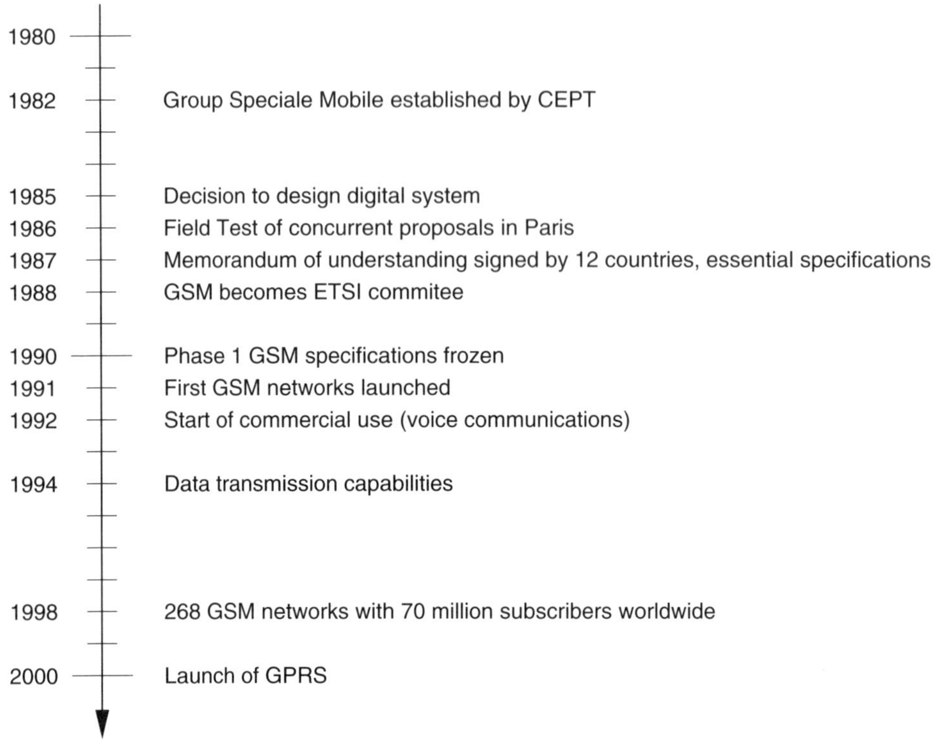

Figure 2.1: Milestones in the GSM history

2.1.2 Specifications and Phases

Initially it was planned to introduce three phases of development

1. specification writing,
2. specification validation and
3. field test with operational equipment.

Since the initial idea of separating the validation phase from the specification writing phase was seen as not practicable, in 1988 it was agreed to start 1991 with a limited service set. Therefore the services were split into two phases, phase 2 being a functional enhancement of phase 1 services that only comprise the most common ones. In 1993 the same reason separated phase 2 into phase 2 and phase 2+ (see Figure 2.2).

In the GSM specifications the functions and interfaces are specified in detail, but not the hardware. They consist of 12 series (see Table 2.1), which have been written by different working parties and a number of expert groups.

The specifications have later been extended to include an air interface also for the 1800 MHz frequency range, GSM 1800.

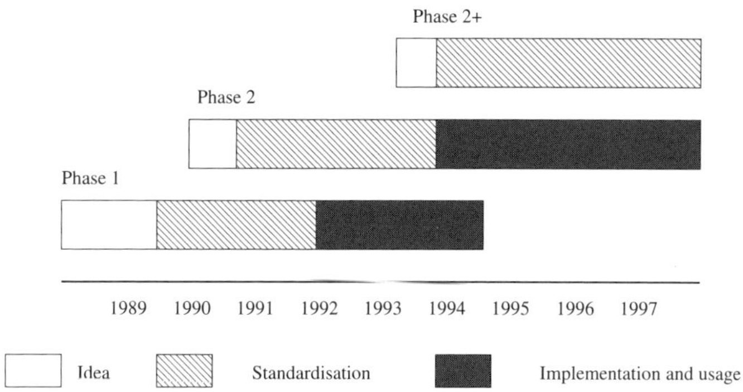

Figure 2.2: Time schedule of GSM phases 1 to 2+

Table 2.1: The series of GSM recommendations

Series	Content
00	Preamble
01	General aspects, terminology and service introduction phases of the GSM *Public Land Mobile Network* (PLMN)
02	Definition of telecommunications services, technical aspects concerning tariffs and international billing procedures
03	Definition of network functions such as traffic routing, handover, security issues relating to network access, network planning
04	Description and definition of protocols and interfaces between *Mobile Station* (MS) and *Base Station* (BS)
05	Radio path functions such as multiplexing, channel coding, synchronization and interleaving
06	Speech processing and speech coding functions
07	Adaptation of terminal equipment and transmission rates
08	Description of interface functions between *Base Station Subsystem* (BSS) and *Mobile-services Switching Center* (MSC)
09	Definition of *Interworking Function* (IWF) between one or more GSM networks and different fixed networks
11	Equipment specifications and type approval guidelines
12	Operation and maintenance of a GSM network

2.1.3 Logical Architecture

GSM is basically divided into the *Radio Subsystem* (RSS), the *Network and Switching Subsystem* (NSS) and the *Operation Subsystem* (OSS). Each of these systems contains a number of functional units, where all system functions are realized. The functional units are implemented in various equipment (hardware).

The RSS comprises the *Mobile Station* (MS) and the *Base Station Subsystem* (BSS).

The BSS is realized as a network of neighboring radio cells together providing a complete coverage of the service area. Each cell has a *Base Transceiver Station* (BTS) operating on a set of radio channels, which are different from those used in neighboring cells to avoid interference.

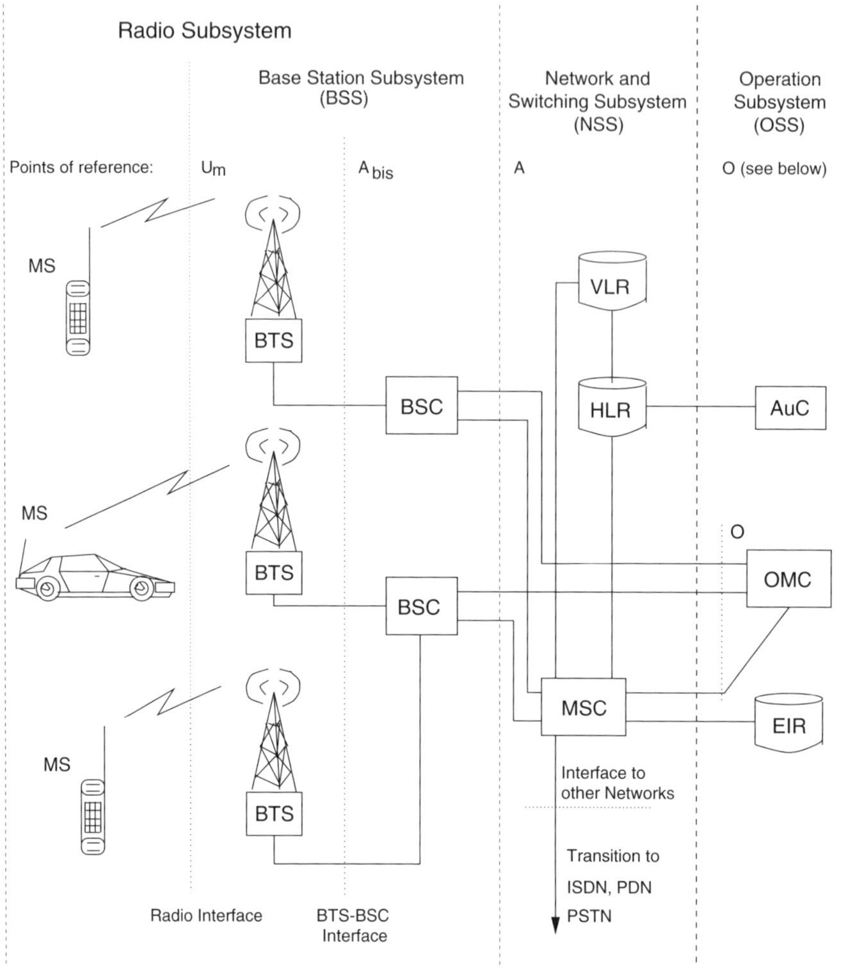

Figure 2.3: The GSM network architecture

A couple of BTSs are managed by one *Base Station Controller* (BSC) handling functions like *handover* or *power control*. Serving a number of BSCs is handled by one MSC that controls calls to and from other networks.

Some databases are part of the NSS. The above mentioned units are all involved in carrying speech or data connections between a mobile station (MS) and other network equipment. One problem arises while trying to establish an incoming external call, terminated by a mobile station. The originator does not know the actual location of the MS. A number of databases is necessary for the network to keep track of the MS and continuously update the records of the mobile.

The following databases are available for call control and network management.

- *Home Location Register* (HLR)
- *Visitor Location Register* (VLR)
- *Authentication Center* (AuC)

- *Equipment Identity Register* (EIR)

The most important one of these databases is the HLR. When acquiring a new subscription from one GSM operator, the subscriber's dataset is stored in the HLR. It contains subscriber information, such as *supplementary services* and authentication parameters. Furthermore information about the location of the MS is stored, i.e., the actual MSC identification the MS is associated with. This information may change while the MS is moving around.

A unit called AuC is connected to the HLR providing information about authentication parameters, ciphering keys and all functions for security purposes.

The VLR contains information about all MSs in this MSC area that are switched on. As soon as an MS *roams* into a new MSC area, the VLR connected to that MSC will request data about the MS from the HLR. At the same time the HLR will be informed about the MSC area holding the MS. If the MS wants to establish a call, the VLR has all the information necessary for call setup without the need for interrogation of the HLR. The VLR can be seen as a distributed HLR.

The *Operating and Maintenance Center* (OMC) is connected to all equipment in the NSS and the BSC. The OMC centrally monitors and controls the network elements and guarantees the best possible service quality for a network. It relies on services of the network management and control functions allocated to the network elements by the hierarchical network management system (*Telecommunications Management Network* (TMN), see [50]).

2.1.4 GSM Mobile Station

A GSM MS consists of two parts. The first part contains all the hardware and software components relating to the radio interface; the second part, known as the *Subscriber Identity Module* (SIM), stores all the subscriber's personal data (see Figure 2.4). The SIM is either installed into the terminal or provided as a *smart card*, which has the function of a key. Once it has been removed from a device, it can only be used for emergency calls, if the network so allows. A mobile subscriber can use the SIM to identify himself over any mobile station in the network, and accordingly a mobile phone can be personalized using the SIM. In addition, each mobile station has its mobile *Equipment Identity* (EI).

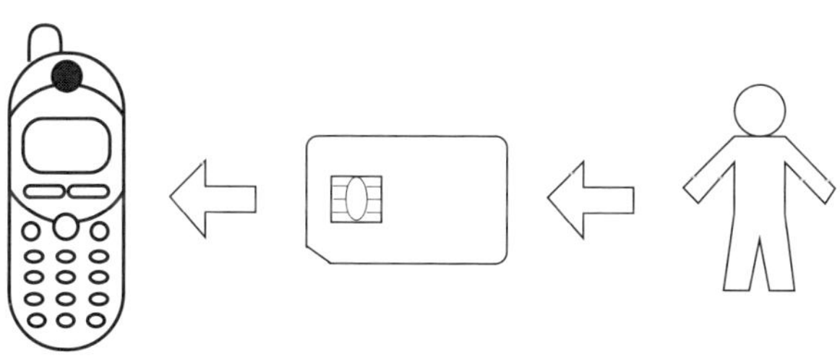

Figure 2.4: Mobile station and SIM card

2.1.5 GSM Numbers and Addresses

The following numbers and identities are assigned for the administration of each mobile station within a GSM network.

- *International Mobile Subscriber Identity* (IMSI)
- *Temporary Mobile Subscriber Identity* (TMSI)
- *Mobile Station international ISDN number* (MSISDN)
- *Mobile Station Roaming Number* (MSRN)

The numbering in GSM follows the rules of ITU-T Rec. E.164 for ISDN. In addition to an MSISDN telephone number, a mobile subscriber has another number—an IMSI with a maximum length of 15 digits that provides him with unique identification.

A temporary address, the *Mobile Station Roaming Number* (MSRN), provides the link to the mobile subscriber's current location. An MSRN is linked to location, is temporarily allocated by a VLR (and stored in the HLR) at the request of the MSC, and is required to make a connection from the fixed network to a mobile station over the *Gateway Mobile-services Switching Center* (GMSC). It consists of the country code of the visited mobile network (*Visited Country Code*, VCC), the local code (area that the user is currently visiting (*Visited NDC*, VNDC)), identification of the *Visited MSC* (VMSC) and the *Visited Subscriber Number* (VSN) allocated by the VLR. The MSRN protects the identity and the location of the mobile user from being discovered through eavesdropping on the signaling traffic.

If a mobile station wishes to send a signal in the network, it uses a local temporary number for identification, a so-called *Temporary Mobile Subscriber Identity* (TMSI), to conceal the IMSI. A TMSI only has local significance and is only valid within the area administered by a VLR. A TMSI is periodically changed to ensure the confidentiality of the mobile subscriber's transmitted information.

2.1.6 Basic Telecommunication Services

The speech- and data-services supported by GSM can be divided into the three categories [24]:

- teleservices,
- bearer services and
- supplementary services.

They will be described in the following sections.

2.1.6.1 Teleservices

A teleservice is a telecommunications service that allows applications-oriented communication between a mobile user and a second user in accordance with standardized protocols. This requires protocols from all seven layers of the ISO/OSI model. The GSM recommendations intend that the teleservice will make use of only one or in any case a small number of different bearer services. Two important characteristics of the

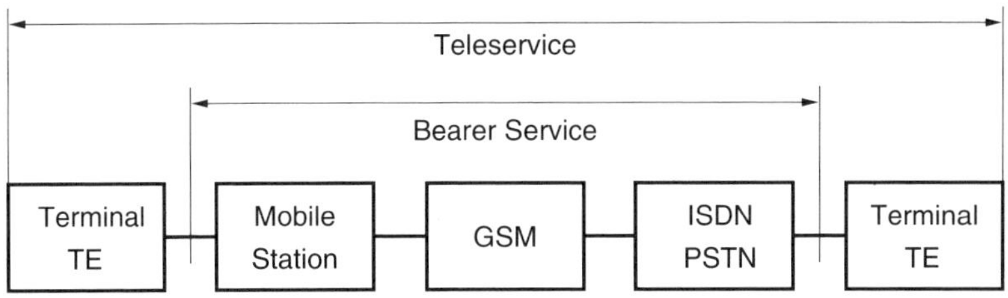

Figure 2.5: Relationship between bearer services and teleservices in a point-to-point connection in GSM

teleservices are system-optimized coded voice transmission and data transmission in which the use of special adapters ensures compatibility with terminal equipment in ISDN or over a modem to the *Public Switched Telephony Network* (PSTN).

The following teleservices are standardized [23].

- Telephone service
- Emergency call service
- Three videotex access profiles
- Telefax Group 3
- Access to electronic mail
- Three short-message services:
  - Short-Message Mobile-Terminated Point-to-Point
  - Short-Message Mobile-Originated Point-to-Point
  - Short-Message Mobile Cell Broadcast.

2.1.6.2 Bearer Services

Bearer services are all telecommunications services that allow mobile subscribers to transmit signals in bit-transparent mode between the mobile subscriber and a user of any partner network [23]. They only offer pure transport services as defined for the lower three layers of the ISO/OSI reference model, namely connection-oriented channel and packet-switched data transmission. The relationship between bearer services and teleservices is outlined in Figure 2.5.

GSM supports different variants of data transmission, thus functioning as a *bearer* for data transmission. The bearer services allow GSM to be used for non-voice services up to 9600 bit/s (see Table 2.2). Because voice transmission is expected to represent the main traffic component and it is projected that data services will equate to less than 10 % of the overall transmission volume, only some of the bearer services were classified as *essential.*

Besides asynchronous and synchronous circuit-switched data services, packet services are provided. They are realized either as asynchronous access to *Packet Assembly Disassembly* (PAD) or as direct synchronous *Packet Access.*

The GSM bearer services for GSM data transfer are offered in transparent (T) and non-transparent (NT) mode. While in the transparent mode a circuit-switched

Table 2.2: GSM bearer services

Service	Structure	Data Rate [kbit/s]	Mode
Data	Asynch.	0.3/1.2/2.4/4.8/9.6	T or NT
Data	Synch.	1.2	T
Data	Synch.	2.4/4.8/9.6	T or NT
PAD	Asynch	0.3/1.2/2.4/4.8/9.6	T or NT
Packet	Synch.	2.4/4.8/9.6	NT
Alternate Speech/Data		13/9.6	
Speech followed by data		13/9.6	

T/NT — Transparent / Non-Transparent
PAD — Packet Assembler Disassembler
Asynchr/Synchr — Asynchronous / Synchronous
PAD — Packet Assembler/Disassembler

connection between the MS and the MSC is provided protected by *Forward Error Correction* (FEC), in non-transparent mode the *Radio Link Control* (RLC) protocol is applied that includes an *Automatic Repeat Request* (ARQ) procedure for the retransmission of erroneous blocks.

Voice services are realized with the bearer services *alternate speech/data* or *speech followed by data.*

2.1.6.3 Supplementary Services

The supplementary services like *subscriber identification*, *call rerouting*, *call forwarding*, etc., are further facilities that are offered to subscribers in addition to the services mentioned above [28]. These are not independent services, and are always offered in conjunction with the teleservices or the bearer services.

2.1.6.4 Support for Value-added Services

Value-added services like *Tariff Area Indication*, *e-mail Access*, *Electronic Banking* or *Travel Information* are not standardized in the GSM specifications. GSM provides dedicated mechanisms to facilitate the provision of value-added services like *Unstructured Supplementary Service Data* (USSD) [12] and *Customized Applications for Mobile network Enhanced Logic* (CAMEL) [6]. As it is up to the respective network operator to provide that type of service, the term "operator-specific" services is used as well.

2.1.7 Radio Interface

The radio interface located between the MS and the BTS influences entirely the spectral efficiency and plays therefore a key factor for operators of cellular systems. The cell structure may vary from several hundred meters radius in an urban environment up to 35 km in a rural one.

Beside the user's speech or data information, additional signaling due to special services or the mobile character of the system has to be accommodated on the radio interface.

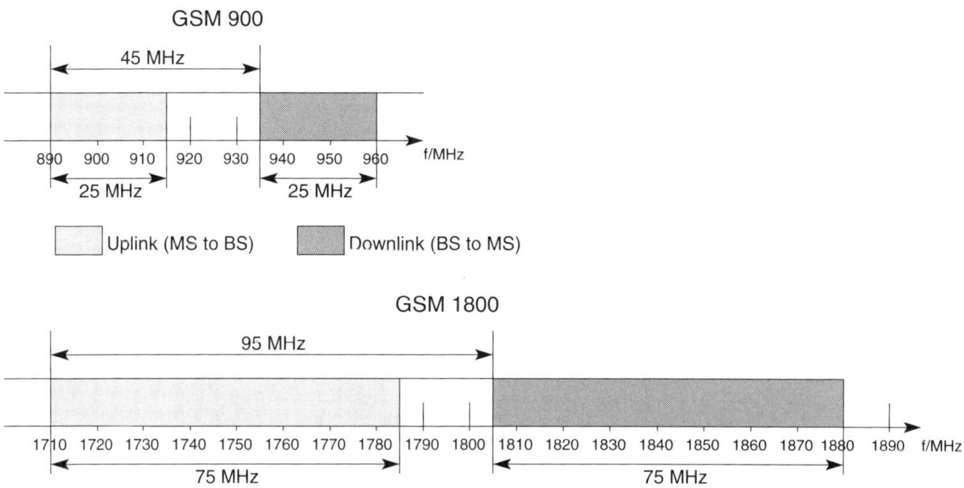

Figure 2.6: Frequency range in GSM 900 and 1800

2.1.7.1 Multiple Access Scheme

Since the radio spectrum is a limited resource shared by all users of a PLMN the radio channel organization method chosen for GSM is a combination of TDMA and *Frequency Division Multiple Access* (FDMA) in combination with different interference reduction techniques (e.g., emission power control and optimized handover decision methods).

Frequency Division Multiple Access (FDMA) In Europe for the GSM system two 25 MHz frequency bands have been reserved around the 900 MHz basis band (see Figure 2.6). The range 890–915 MHz is used for uplink transfer between MS and BS and 935–960 respectively for the downlink direction so that there exists a duplex distance of 45 MHz. They are divided by frequency into 124 carriers each, spaced 200 kHz apart (see Figure 2.7). The different carrier frequencies can be calculated by the following equations :

$$F_u(n) = 890.2\,\text{MHz} + 0.2 \cdot (n-1)\,\text{MHz with } 1 \le n \le 124 \tag{2.1}$$

$$F_d(n) = 935.2\,\text{MHz} + 0.2 \cdot (n-1)\,\text{MHz with } 1 \le n \le 124 \tag{2.2}$$

The *Digital Cellular System* (DCS) 1800 standard developed by ETSI is based on the GSM recommendations but applied to the higher frequency range at 1800 MHz. This is the reason why nowadays DCS 1800 is termed GSM 1800. The frequency ranges assigned in Europe are 1710–1785 MHz (uplink) and 1805–1880 MHz (downlink). A total of 374 carrier frequencies are therefore available to the system:

$$F_u(n) = 1710\,\text{MHz} + 0.2 \cdot (n-1)\,\text{MHz with } 1 \le n \le 374 \tag{2.3}$$

$$F_d(n) = 1805\,\text{MHz} + 0.2 \cdot (n-1)\,\text{MHz with } 1 \le n \le 374 \tag{2.4}$$

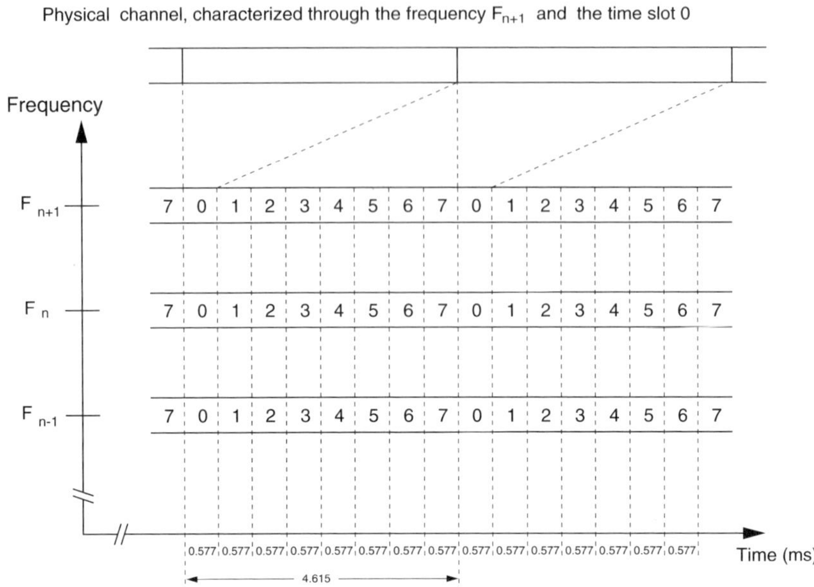

Figure 2.7: Realization of physical channels using FDM and TDM

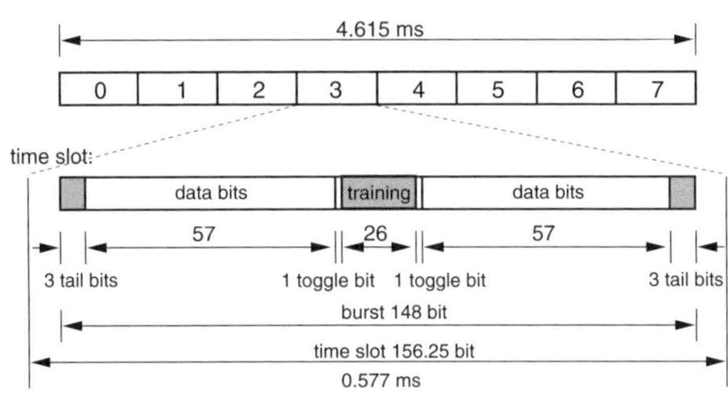

Figure 2.8: TDMA frame structure and normal burst

One or more carrier frequencies are assigned to each BS. The structure and the size of the cells as well as the local environment characteristics and the cell organization determine the carrier frequencies which the BS may reuse, considering the overall interference level. Additionally the network operator can use the optional frequency hopping capability which is implemented on all MSs. The advantage is that the quality on all communications is averaged through interferer's diversity.

Time Division Multiple Access (TDMA) Each of the carrier frequencies is divided into eight time slots (see Figure 2.7). These form a TDMA frame so that each physical channel may be described by a time slot number and a carrier frequency. Each TDMA frame is 4.614 ms long.

Table 2.3: Time parameters of multiple TDMA frames

TDMA Frame Type	Duration	Notes
Bit	$\frac{48}{13}\,\mu s \approx 3.69\,\mu s$	smallest unit
Time slot	$\frac{15}{26}\,ms \approx 0.577\,ms$	156.25 bit
TDMA-frame	$\frac{120}{26}\,ms \approx 4.615\,ms$	8 slots
26-frame	$120\,ms$	26 TDMA frames
51-frame	$\frac{3060}{13}\,ms \approx 235.385\,ms$	51 TDMA frames
Superframe	$6.12\,s$	1326 TDMA frames
Hyperframe	$3\,h \quad 28\,min \quad 53\,s \quad 760\,ms$	2 048 superframes

Within a time slot data are transmitted in the form of bursts. One time slot lasts for 577 μs (15/26 ms), that is 156.25 bit durations. In order to fulfill different tasks there exist four types of bursts:

- Normal burst.
- Frequency correction burst.
- Synchronization burst.
- Access burst.

Figure 2.8 depicts the structure of a normal burst within a TDMA frame.

2.1.7.2 Channel Organization

In GSM physical and logical channels are distinguished. One time slot of a TDMA frame is referred to as a physical channel. On a physical channel a great variety of information is transmitted. Different logical channels are mapped onto a physical channel and they are defined by the number and position of their corresponding burst periods. Logical channels can only be deployed in certain combinations and on certain physical channels. Different channel combinations have been defined, which are realized and offered by the BS [111]

The definition is cyclic and the entire pattern of recurrence is attached to hyper-, super- and multiframes (see Figure 2.9). Logical channels are separated into traffic and control channels. These dependencies and some on top of them are set out in Table 2.3.

2.1.7.3 Traffic Channels

A *Traffic Channel* (TCH) is used to carry speech or circuit-switched data traffic. TCHs are defined using a 26-frame multiframe, or a group of 26 TDMA frames. The length of a 26-frame multiframe is 120 ms. TCHs in up- and downlink are separated in time by three burst periods (time slot duration), so that the MS does not have to transmit and receive simultaneously.

TCHs are further distinguished by fullrate and halfrate TCHs (TCH/F and TCH/H). Halfrate channels for speech assume an appropriate speech codec (see Ta-

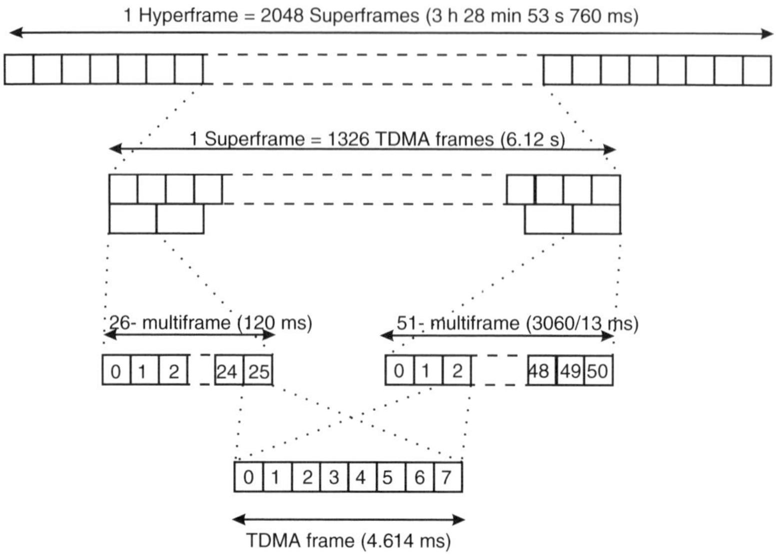

Figure 2.9: Hyper-, super- and multiframe structure

ble 2.4). Data transmission on a TCH is limited to 9.6 kbit/s and 4.8 kbit/s respectively.

2.1.7.4 Signaling Channels

User data is not the only flow of information to be transported during a call. Signaling messages must also be conveyed to allow the mobile station and the network management to arrange parameters related to the user, the network or aspects of communication. Even if no communication is in progress it is necessary to exchange information, e.g., for location updating while roaming.

The pool of signaling channels can be divided into the three major groups (see Table 2.4)

- *Broadcast Control Channels* (BCCHs)
- *Common Control Channels* (CCCHs)
- *Dedicated Control Channels* (DCCHs).

Broadcast Control Channel (BCCH) The BCCH is a point-to-multipoint, unidirectional control channel from the fixed subsystem to the mobile station which is intended to broadcast a variety of information, including information necessary for the MS to register in the system.

Included subchannels are a *Frequency Correction Channel* (FCCH), which is used to allow an MS to accurately tune to a BS, and a *Synchronization Channel* (SCH), which is used to provide TDMA frame-oriented synchronization data to an MS.

On the FCCH a sine wave signal is transmitted for two purposes. On the one hand it has to make sure that this is really the BCCH carrier and on the other hand the receiving mobile needs to have the possibility to synchronize to the frequency.

Table 2.4: GSM logical channels

Group	Channel Name	Function	Direction
Traffic Channel (TCH)	TCH/F	Full Rate TCH	UL/DL
	TCH/H	Half Rate TCH	UL/DL
Broadcast Channel	BCCH	Broadcast Control	DL
	FCCH	Frequency Correction	DL
	SCH	Synchronization	DL
Common Control Channel (CCCH)	RACH	Random Access	UL
	AGCH	Access Grant	DL
	PCH	Paging	DL
	NCH	Notification	DL
Dedicated Control Channel (DCCH)	SDCCH	Stand-alone Dedicated Control	UL/DL
	SACCH	Slow Associated Control	UL/DL
	FACCH	Fast Associated Control	UL/DL

For synchronizing to the structure within a certain cell the SCH is used. It is necessary to decide if the chosen base station is really a GSM BS, to receive information about the TDMA frame structure in this cell and to store the *Base Station Identification Code* (BSIC) of the chosen base station. This BSIC can only be decoded if the base station belongs to a GSM network.

Common Control Channel (CCCH) This is a point-to-multipoint, bidirectional control channel, which is first of all intended to carry signaling information necessary for access management functions, like allocation of dedicated control channels. Three kinds of subchannels can be included in a CCCH. The *Paging Channel* (PCH) is a channel on the downlink used for paging mobile stations, a *Random Access Channel* (RACH) is used for requesting the assignment of a DCCH and an *Access Grant Channel* (AGCH) is used to assign an MS to a specific DCCH.

Within specific time intervals the MS will listen to the PCH to see if the network wants to get in contact with the MS. The reason could be an incoming call or an incoming *short message*. The information on the PCH is a paging message including the IMSI or a TMSI.

If listening to the PCH the MS will realize that it is being paged. The mobile station answers, requesting a signaling channel, on the RACH. This channel is also used by the MS if it wants to get in contact with the network e.g. for call setup.

Network initiated signaling channel assignment is performed on the AGCH in the downlink direction as point-to-point.

Dedicated Control Channel (DCCH) Divided into three types of subchannels, this point-to-point control channel provides access to the *Slow Associated dedicated Control Channel* (SACCH) and the *Stand-alone Dedicated Control Channel* (SDCCH) and *Fast Associated dedicated Control Channel* (FACCH).

The allocation of an SDCCH is not linked to the allocation of a TCH but it is used before assigning the MS to a TCH for providing authentication to an MS, location

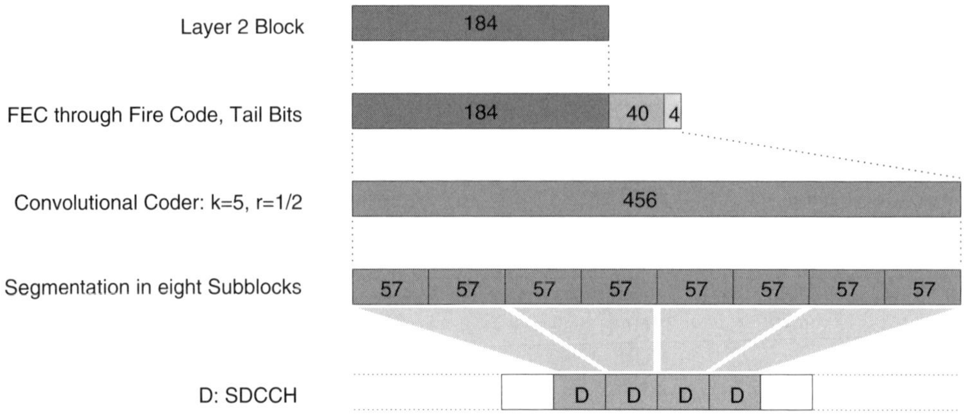

Figure 2.10: Coding and interleaving in an SDCCH

updates and assignment to a TCH. In contrast to the SDCCH the *Associated Control Channel* (ACCH) is linked to the allocation of a TCH.

Among others the call setup procedure is performed on the SDCCH as well as the transmission of text messages like SMS and *cell broadcast* in idle mode. Its direction is both up- and downlink in a point-to-point connection.

Within specific time intervals on the SACCH information is transmitted. On the uplink the MS sends averaged measurements about signal strength and quality on its own and neighboring base stations. Received information on the downlink carries values about transmission power to be used by the MS and also an instruction on the *timing advance* (see [111]).

If during the conversation a handover must be performed the FACCH is used. As the name FACCH indicates this channel is a *fast* channel and has a maximum transfer rate of 9.2 kbit/s. The needed slots are stolen from the slots of the traffic channel.

2.1.7.5 Channel Coding and Modulation

Different channel coding methods are applied on the various channels for the different services. A Fire code—a linear, binary block code derived from the cyclic block code family—is used with most signaling channels. This code has a generator polynomial that offers good error detection and/or error correction when errors occur in groups (*burst errors*). Forty Fire code parity bits are added to the 184 data bits of a layer-2 *Protocol Data Unit* (PDU) (see Figure 2.10). This allows groups of errors of up to 11 bits to be detected and corrected.

All channels have *Forward Error Correction* (FEC) protection through convolutional coding. A convolutional code with the ratio $r = 1/2$ and the influence length $k = 5$ (see Figure 2.10) is used on the control channels. Interleaving allows correlated errors caused by a short signal fading to appear as single errors at the decoder of the receiver.

Two types of interleaving are in use, bit interleaving and block interleaving. The bit interleaving with a depth of n is based on an algorithm, where the 456 information bits of a convolutional coded layer-2 block are arranged into rows mod n in a table [111]. The bits are read out for transmission from the table row by row.

The block interleaving procedure further contributes to spread originally neighboring bits during the radio transmission to improve the resistance against the so-called error bursts resulting from signal fading. On the FACCHs block diagonal interleaving occurs in eight half bursts. On all other control channels (with the exception of RACH) interleaving is in four whole bursts.

Depending on the channel types these coded data are encapsulated into the related bursts (see Figure 2.8), which are transmitted with a gross bit rate (modulation bit rate) of 270.8 kbit/s.

The GSM system uses a digital 0.3 GMSK modulation format. GMSK stands for *Gaussian Minimum Shift Keying*, in which *minimum shift* means that there is no discontinuity between the phase of the carrier and the time. The modulation format 0.3 GMSK means that the modulated signal passes through a Gauss filter with the product of the 3 dB bandwidth B and the bit period T producing the value $BT = 0.3$ [50].

2.2 GPRS—General Packet Radio Service

The main goal of integrating the *General Packet Radio Service* (GPRS) into the GSM is to use the GSM radio resources more efficiently than is possible with the existing GSM Phase 2 services. This is realized by allowing a number of logical connections per bearer to co-exist and by multiplexing packet data of these connections onto the GSM physical channels.

GPRS has been standardized by the ETSI as part of the GSM Phase 2+ development (see Section 2.1.2). The phase 2+ specifications are defining the implementation of packet switching within GSM, which is essentially a circuit-switched (CS) technology.

Packet-switching means that GPRS radio resources are used only when users are actually sending or receiving data. Rather than dedicating a radio channel to a mobile data user for a fixed period of time, the available radio resource can be shared between several users. The actual number of users supported depends on the application being used and how much data is being transferred. Through multiplexing of several logical connections to one or more GSM physical channels, GPRS reaches a flexible use of channel capacity for applications with variable bit rates.

Additionally it brings *Internet Protocol* (IP) capability to the GSM network and enables the access to a wide range of public and private data networks using industry standard data protocols such as *Transport Control Protocol* (TCP)/*Internet Protocol* (IP) and X.25.

The standardization of the GPRS was broadly concluded in 1997. However, details were regularly discussed in the boards of standardization and the GPRS standard was regularly modified. First proposals for a packed-oriented data service in GSM were published in 1991 [113, 46, 47, 112, 42].

The first GPRS-based services have been available since 2001 in Europe. Many countries worldwide will introduce GPRS in the next few years. With these new services mobile multimedia applications with net bit rates of up to 117 kbit/s will be offered and established on the market.

2.2.1 Design Approach

The motivation to define a GSM packet-switched data service in 1994 was motivated by the growing number of dedicated packet radio networks. The provision of a circuit-switched channel for bursty traffic was felt to be inefficient. Applications in mind were fleet management, logistics, telematics and mobile offices [68].

In order to realize economically priced packet services, it was the premise to change the GSM components as little as possible and to develop the new service in consideration of the limits of the existing GSM tele- and bearer services (see Section 2.1.6).

GPRS integrates a packet-based air interface into the existing circuit-switched GSM network. The GSM infrastructure is not to be replaced. A couple of new network elements have been added (see Section 2.2.2).

The GPRS specification does not provide an upper limit for the amount of data which can be transmitted per access. GPRS was initially designed for

- the frequent, regular transmission of short data packets up to 500 bytes, and
- the irregular transmission of short to medium-sized data packets up to a few kbytes.

The basic approach to integrate the packet data service into the GSM standard represents the reservation and the logical subdivision of certain GSM channels. The number of channels allocated for GPRS can be dynamically adapted to the workload situation in the respective radio cell.

2.2.2 Logical Architecture

The existing GSM network does not provide sufficient functionality to realize a packet data service. Integrating GPRS into a GSM network requires the addition of components which provide the packet-switched service (see Figure 2.11). Hence, the GSM network is extended by two additional nodes:

Gateway GPRS Support Node (GGSN): The GGSN serves as the interface towards external *Packet Data Networks* (PDNs) or other *Public Land Mobile Networks* (PLMNs). Here, switching functions are fulfilled, e.g., the evaluation of *Packet Data Protocol* (PDP) addresses and the routing to mobile subscribers via the SGSN.

Serving GPRS Support Node (SGSN): The SGSN represents the GPRS switching center in analogy to the MSC. Packet data addresses are evaluated and mapped onto the IMSI. The SGSN is responsible for the routing inside the packet radio network and for mobility and resource management. Furthermore, it provides authentication and encryption for the GPRS subscribers.

For the communication between the SGSN and the GGSN within one PLMN, the Intra PLMN *IP Version 6* (IPv6) or *IP Version 4* (IPv4) backbone is used. SGSN and GGSN encapsulate and decapsulate, respectively, the packets using a special protocol called the *GPRS Tunneling Protocol* (GTP) that operates on top of standard TCP/IP protocols. The SGSN and GGSN functions may also be combined into one physical node.

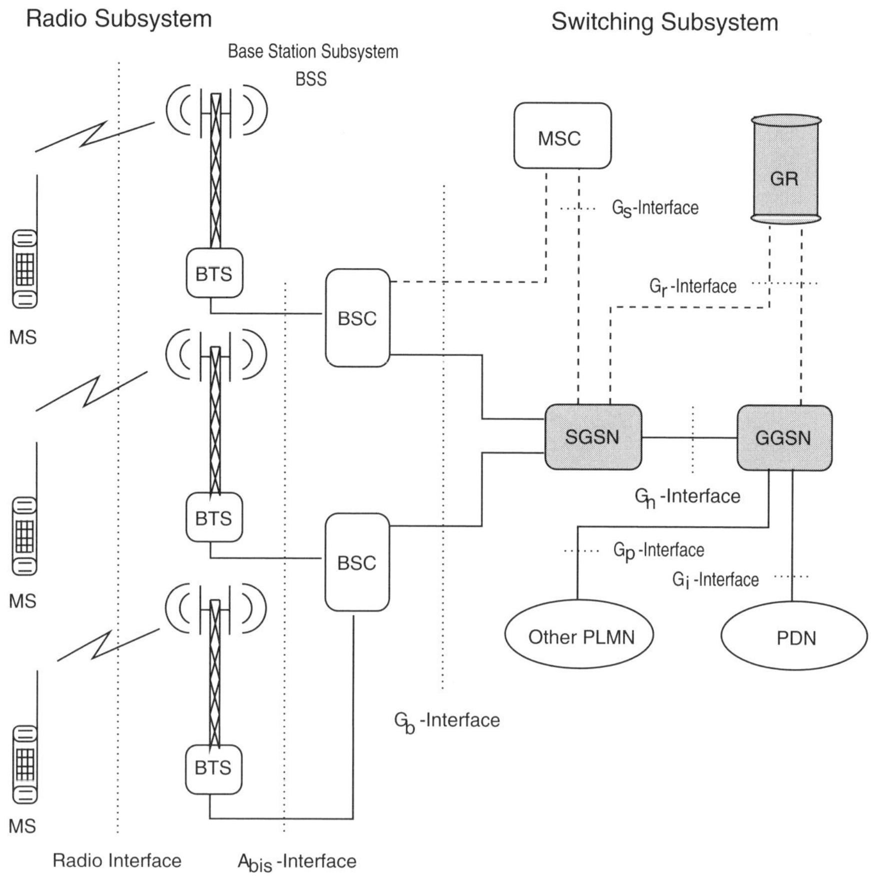

Figure 2.11: GPRS logical architecture

With the extension of the existing GSM network by GPRS, interfaces and reference points had to be redefined. The defined interfaces are displayed in Figure 2.12. The dotted lines allude to signaling traffic between the corresponding elements. Solid lines mean that user data can be transmitted additionally at these reference points.

In GPRS the proper data traffic is handled by the SGSN. The MSC serves only for signaling.

2.2.3 Service Types

A PLMN provider is responsible for the data transfer between the GPRS *Service Access Point* (SAP) in the fixed network and the GPRS SAP in the MS. Two service categories are specified [25]:

Point-To-Point (PTP) Using the PTP service single packets can be transmitted between two subscribers. The PTP service is offered both in a connection-oriented mode *Connection Oriented Network Service* (CONS) and in a connectionless mode *Connectionless Network Service* (CLNS). Applications using the PTP service can be classified regarding their communication properties into:

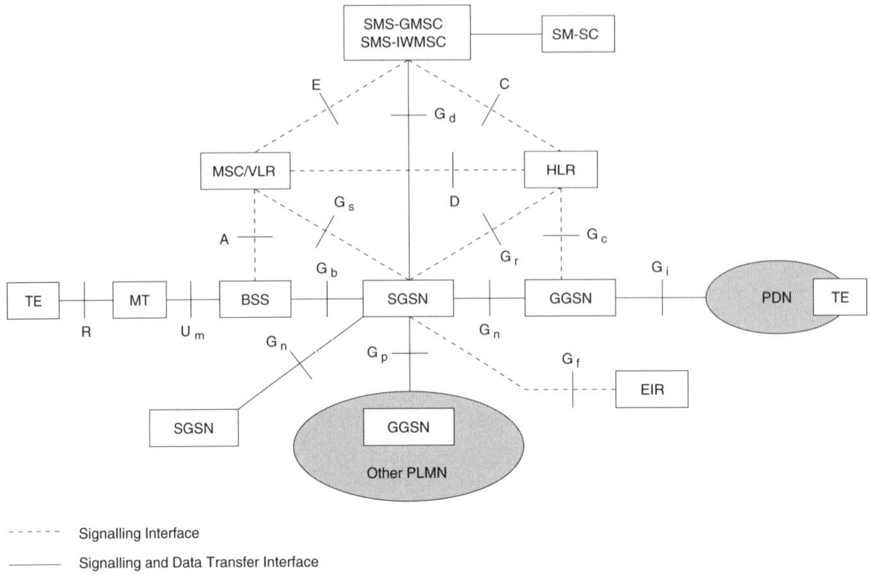

Figure 2.12: GPRS interfaces and reference points

Table 2.5: Relationship between service type and service originator/receiver

Packet flow	Point-to-Point		Point-to-Multipoint	
	PTP-CONS	PTP-CLNS	PTM-M	PTM-G
fixed → mobile	×	×	×	×
mobile → mobile	×	×	×	×
mobile → fixed	×	×	–	×

non-interactive There is no dependence between the single data packets.

interactive or dialog-oriented There is a logical relationship between the service subscribers for a particular period. This period can range from several seconds to several hours.

Point-To-Multipoint (PTM) This service supports the transmission of data packets between a service user and a specified group inside a certain geographical region. It is divided into:

(PTM-M) *PTM Multicast* (PTM-M) comprises calls which are broadcast in the area specified by the call initiator. Either all subscribers in this region or the members of a group are addressed. In the newest technical specifications this service is realized by IP Multicast (IP-M).

(PTM-G) The messages are addressed explicitly to a specified group and are sent in all regions where group members are located.

Table 2.5 shows that services except the multicast service can be originated both from the mobile radio network *Mobile Originated Call* (MO-C) and from the fixed network *Mobile Terminated Call* (MT-C).

2.2.4 Parallel Use of Services

During a GPRS session circuit-switched services (speech as well as data) may still be initiated and used. Similarly, it is possible to send and receive GPRS data while carrying out a telephone call. Parallel use of these services is provided for PTP as well as PTM services; however, it causes fluctuating transfer rates according to traffic load and affects the QoS.

Furthermore, it is possible to additionally utilize the SMS, for MO-Cs as well as MT-Cs. An SMS message may, however, be transmitted with delay or a lower data rate, according to the current load situation. An SMS cell broadcast message may be sent in parallel to GPRS services but is not allowed to be used simultaneously to a circuit-switched connection.

Three equipment and subscriber classes have been proposed to offer the user a packet-data service with various performance characteristics.

Subscriber class A Simultaneous usage of all services with unchanging speech quality.

Subscriber class B Restricted simultaneous usage of services with reduced speech quality.

Subscriber class C No simultaneous usage of services. For subscriber class C it is possible to receive short messages at any time, however.

2.2.5 Radio Interface

The radio interface U_m is located between MS and BSS. The GSM recommendations combine a *Frequency Division Multiplex* (FDM) and a *Time Division Multiplex* (TDM) scheme to the radio resources as described in Section 2.1.7.1.

The communication over the GPRS radio interface comprises functions of the *Radio Link Control* (RLC)/*Medium Access Control* (MAC) layer and the physical layer. For the transfer of layer-2 messages over the radio interface the GSM logical channel concept is reused (see Section 2.1.7.2).

2.2.5.1 Channel Concept

The basic approach to integrate the packet data service into the GSM standard represents the reservation and the logical subdivision of certain GSM channels for GPRS. The physical channels dedicated to packet data traffic are called *Packet Data Channels* (PDCHs).

Similar to GSM, logical channels are mapped onto physical channels using a cyclically recurring multiframe structure. Two 26-frame multiframes are assembled to one GPRS 52-frame multiframe. A 52-frame multiframe represents one physical GPRS channel consisting of 12 radio blocks and four `Idle` bursts, each block comprising four radio bursts distributed on time slots with the same *Timeslot Number* (TN) in consecutive TDMA frames (see Figure 2.13 and Figure 2.14).

Since one TDMA frame represents eight physical channels to be shared by GSM and GPRS, there have to be strategies on how to distribute these resources, whether fixed or on-demand (see Figure 2.15). In [17] the principles concerning this task are specified. The specific details of implementation are left to manufacturers and network operators.

The master-slave concept One or more PDCHs are run as *Master Packet Data Channels* (MPDCHs) (see Figure 2.14), and provide *Packet Common Control Channels* (PCCCHs) that carry the necessary control and signaling information to initiate a packet data transmission. These PCCCHs have to be run if signaling information is not transmitted via existing *Common Control Channels* (CCCHs).

Furthermore PDCHs operate as slaves and are utilized for transmission of user data over *Packet Data Traffic Channels* (PDTCHs) and associated control over *Packet Associated Control Channels* (PACCHs).

The capacity-on-demand concept GPRS does not require fixed allocated PDCHs. Capacity assignment for packet data transmission can be done according to actual demand. The decision about the number of fixed and on-demand PDCHs is left with the radio network operator.

There are several mechanisms to increase or diminish the number of actually assigned PDCHs on a capacity-on-demand base.

Load monitoring The PDCH utilization is supervised by a load monitor instance that should be implemented as part of the MAC functionality.

Dynamic allocation of PDCHs Unused channels, whether by GSM or GPRS, may be allocated as PDCHs to increase the GPRS QoS. PDCHs are released in order to meet the obligations of higher priority services.

Release of PDCHs Fast release of PDCHs is an important criterion for a pool of radio resources to be dynamically shared by circuit as well as packet switched services. To achieve this goal, there are the following possibilities:

- Channel release is delayed until there are no more packet data flows occurring on the regarded PDCH.
- Each user having allocated the PDCH to be released has to be notified, e.g. by `Packet Resource Reassignment` messages from the network side.

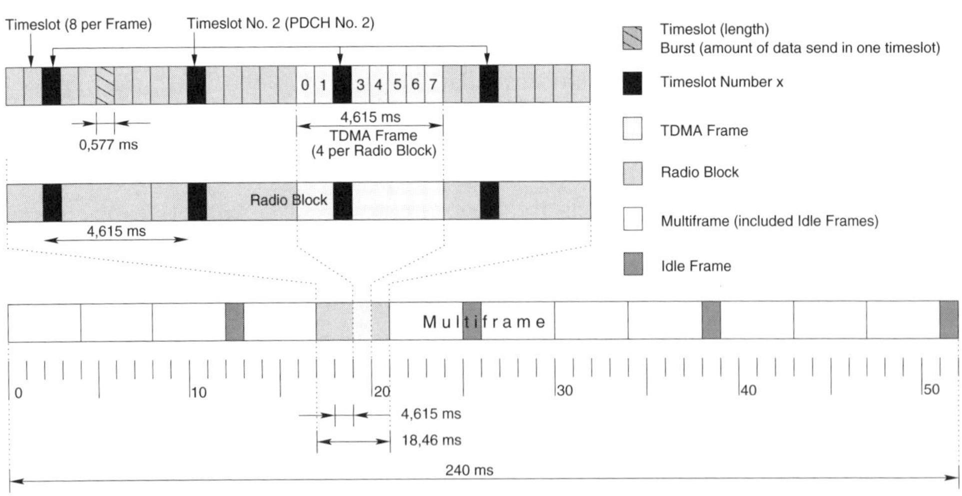

Figure 2.13: Time duration of a time slot, TDMA frame, radio block and 52-frame

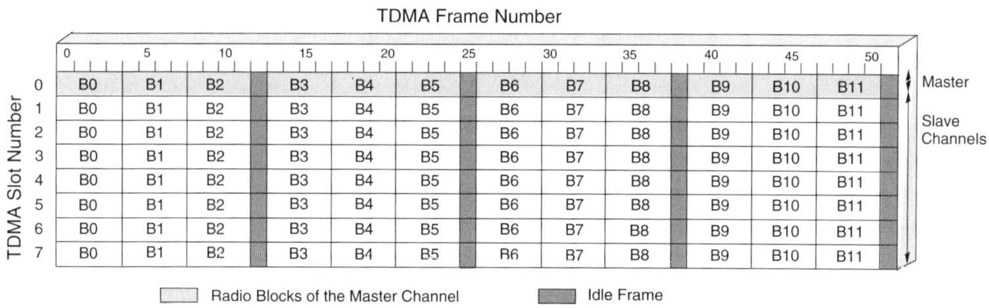

Figure 2.14: GPRS channel structure

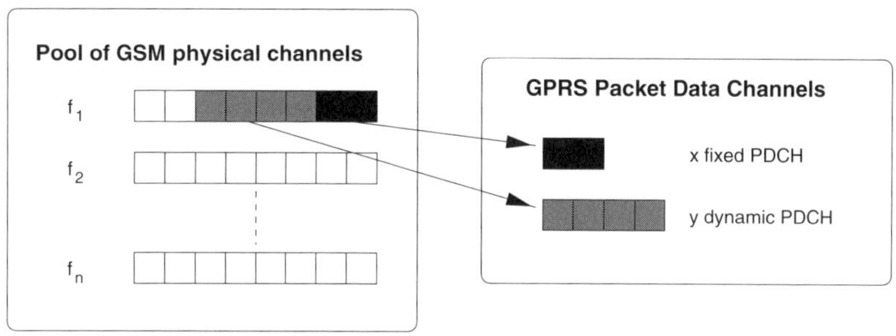

Figure 2.15: Assignment of GSM physical channels for GPRS

These messages have to be sent via PACCHs to each single MS. The PACCHs may be located on different PDCHs.

- The channel release notification is broadcast. There has to be one PDCH monitored by all MSs to assure the reception of this notification. A more effective solution is to broadcast the release notification solely through those PDCHs on which the respective MSs' PACCHs are located.

In practice, a combination of the above methods can be implemented. In case of an MS not receiving a channel release notification there will be temporary channel interference because of the MS using a channel provided for other services. However, the MS will terminate the packet data flow as soon as it receives an inappropriate response on the downlink, and initiate a retransmission on a different PDCH.

2.2.5.2 Logical Channels

The packet data logical channels are mapped onto the physical channels dedicated to packet data (PDCHs).

Table 2.6 lists the GPRS logical channels and their functions. A detailed description for each channel is presented below.

Packet Common Control Channel (PCCCH) The PCCCH comprises the following logical channels for GPRS common control.

- The *Packet Random Access Channel* (PRACH) is used by the MS in the uplink direction to initiate uplink transfer, e.g., for sending data or a paging response.

Table 2.6: GPRS logical channels

Group	Channel	Name	Direction	Function
PCCCH	PRACH	Packet Random Access Channel	UL	random access
	PPCH	Packet Paging Channel	DL	paging
	PAGCH	Packet Access Grant Channel	DL	access grant
	PNCH	Packet Notification Channel	DL	multicast
PBCCH	PBCCH	Packet Broadcast Control Channel	DL	broadcast
PTCH	PDTCH	Packet Data Traffic Channel	UL/DL	data
	PACCH	Packet Associated Control Channel	UL/DL	assoc. control

In other words the channel is used by the MS to initiate a packet transfer or respond to paging messages. On this channel an MS transmits access bursts with long guard times. On receiving access bursts, the BSS assigns a value for the *Timing Advance* (TA) algorithm to each terminal.

- The *Packet Paging Channel* (PPCH) is used in the downlink direction to page an MS prior to downlink packet transfer. The PPCH uses paging groups in order to allow usage of *Discontinuous Reception* (DRX) mode. PPCH can be used for paging packet data services.

- The *Packet Access Grant Channel* (PAGCH) is used in the establishment phase of the packet transfer to send a resource assignment in the downlink direction to an MS requesting a packet transfer. Additionally, resource assignment for a downlink packet transfer can be sent on a PACCH if the MS is currently involved in a packet transfer.

- The *Packet Notification Channel* (PNCH) is used in the downlink direction to send a PTM-M notification to a group of MSs prior to a PTM-M packet transfer. The notification has the form of a resource assignment for the packet transfer. DRX mode will be provided for monitoring the PNCH. Furthermore, a `PTM-M new message` indicator may optionally be sent on all individual paging channels to inform MSs interested in PTM-M when they need to listen to the PNCH.

Packet Broadcast Control Channel (PBCCH) The *Packet Broadcast Control Channel* (PBCCH) broadcasts packet data specific system information. If the PBCCH is not allocated, the BCCH transmits the system information to all GPRS terminals in a radio cell.

Packet Traffic Channel (PTCH) The PTCH is allocated to carry user data and associated control information and is subdivided into the following channels.

- The *Packet Data Traffic Channel* (PDTCH) is a channel allocated for data transfer. It is temporarily dedicated to one MS or in the PTM case to a group of MSs. In multislot operation, one MS may use multiple PDTCHs, e.g., more than one PDTCH, simultaneously for individual packet transfer.

- The *Packet Associated Control Channel* (PACCH) is used to convey control information related to a given MS, such as an acknowledgement message. The

PACCH also carries resource assignment and reassignment messages comprising the assignment of capacity for PDTCHs and for further occurrences of PACCH. One PACCH is associated with one or several PDTCHs concurrently assigned to one MS.

2.2.5.3 Mapping of the Logical Channels

The mapping of the logical channels is defined by the multiframe structure. As described in Section 2.2.5.1 the multiframe structure for a PDCH consists of 52 TDMA frames, divided into 12 blocks (of four frames) and four idle frames according to Figure 2.16. The mapping of the logical channels onto the radio block periods is defined by means of the ordered list of blocks (B0, B6, B3, B9, B1, B7, B4, B10, B2, B8, B5, B11). With this list a certain number of channels per PDCH is uniformly distributed over the multiframe. For example one PBCCH and three PAGCHs on the master channel would be allocated to B0, B6, B3 and B9 of the first PDCH.

PDCHs that contain PCCCHs are indicated on the BCCH. The number of PDCHs containing PCCCHs is defined in the `BS PCC CHANS` parameter. The following restrictions are defined in [16] concerning PCCCHs.

Only one PDCH—the master channel—contains one or more PBCCHs All other channels do not contain PBCCHs and are called slave channels. The logical channel PAGCH can only be mapped on the master channel and the slave channels containing PCCCHs. The response message to a `Packet Channel Request` has to be sent on the same PDCH that was used for the `Packet Channel Request`.

The PRACH can only be mapped on an uplink PDCH containing PCCCHs.

On PRACHs, access bursts are used, while on all other packet data logical channels, radio blocks comprising four normal bursts are used. The only exception are messages on uplink PACCHs, which can comprise four consecutive access bursts to increase robustness. The detailed mapping rules of the single channels as defined in [16] are described in the following (see Figure 2.16).

Downlink

Mapping of PBCCHs: On the downlink of the master channel the first block in the ordered list of blocks (B0) is always used as a PBCCH. If required up to three more blocks on the same PDCH can be used as additional PBCCHs. The parameter `BS PBCCH BLKS` indicates the total number of PBCCH blocks (from one up to four) on the downlink master channel. Therefore on this PDCH the first `BS PBCCH BLKS` blocks in the ordered list of blocks are reserved for PBCCHs.

Any additional PDCH containing PCCCHs is indicated on the PBCCH. On these PDCHs the `BS PBCCH BLKS` first blocks in the ordered list of blocks are used as PDTCHs or PACCHs in the downlink.

Mapping of PAGCHs, PNCHs, PDTCHs and PACCHs: On any PDCH containing PCCCHs (with or without PBCCH), the next `BS PAG BLKS RES` blocks in the ordered list of blocks are used as PAGCHs, PNCHs, PDTCHs or PACCHs. The `BS PAG BLKS RES` parameter is broadcast on the PBCCH and ranges from 0 up to `12 - BS PBCCH BLKS`.

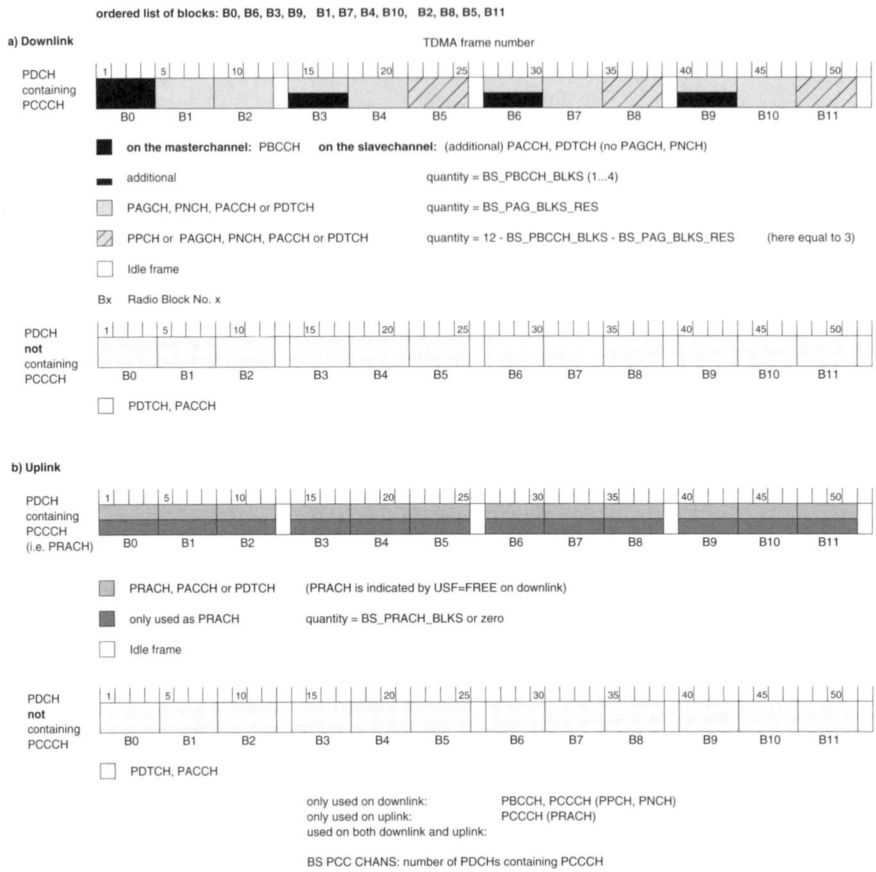

Figure 2.16: Mapping of the logical channels onto the radio blocks

Mapping of PPCHs: The remaining blocks in the ordered list of blocks are used as PPCHs. But they can also be used as PAGCHs, PNCHs, PDTCHs or PACCHs, as in the channel mapping before.

Uplink

Mapping of PRACHs: On a PDCH containing PCCCHs, all blocks in the multiframe can be used as a PRACH—indicated by a an Uplink State Flag in the previous downlink message—or as a PDTCH or a PACCH.

Optionally the `BS PRACH BLKS` first blocks in the ordered list of blocks are only used as PRACHs. The `BS PRACH BLKS` parameter is broadcast on the PBCCH. The remaining blocks in the multiframe are used as PRACH, PDTCH or PACCH.

Mapping of PACCHs and PDTCHs: On a PDCH that does not contain a PCCCH, i.e., that does not contain a PRACH (`BS PRACH BLKS`=0), all blocks are used as PDTCHs or PACCHs.

3
Data Transmission in GPRS Networks

The GPRS protocol architecture follows the ISO/OSI reference model. The GPRS user plane, also called transmission plane, comprises all protocols used for user data transmission. In GPRS networks the user plane can be considered separate from the control plane, since communication is ongoing in two phases. While a so-called *Packet Data Protocol* (PDP) context, a logical context between an MS and an external *Packet Data Network* (PDN) based on TCP/IP or X.25, can be set up without actually transmitting user data, the user plane will only be utilized when the MS is actually transmitting or receiving user data. Nevertheless some protocols of the user plane are reused to carry signaling messages (see Chapter 4).

The kernel of the GPRS standard is the *Radio Link Control* (RLC)/*Medium Access Control* (MAC) protocol that operates on top of the GSM channel structure (see Section 2.2.5.1) and realizes the reliable packet-oriented transmission between *Mobile Station* (MS) and *Base Station* (BS) by statistical multiplexing of several logical connections on the *Packet Data Channels* (PDCHs) available for GPRS in the radio cell. While RLC/MAC contains already an ARQ scheme to handle transmission errors on the air interface, the *Logical Link Control* (LLC) protocol provides a reliable ciphered link between MS and *Serving GPRS Support Node* (SGSN) applying flow control and error handling mechanisms known from link layer protocols in fixed networks like ISDN. The ARQ scheme in LLC is not designed to handle transmission errors on the air interface, but for handling frame losses and errors in the fixed network or for link recovery after a cell change.

The *Base Station Subsystem GPRS Protocol* (BSSGP), which is operating on top of a network service such as *Frame Relay* (FR), is responsible for flow control between BSS and SGSN. To support different network layer standards that may be used in GPRS networks the *Sub-Network Dependent Convergence Protocol* (SNDCP) adapts the network layer PDUs to the GPRS logical link. Finally the *GPRS Tunneling Protocol* (GTP) is used as part of the GPRS user plane to realize a transport service for IP datagrams by establishing an IP tunnel through the GPRS core network between SGSN and GGSN.

This chapter will describe all GPRS-specific protocols (see shaded blocks in Figure 3.1) in detail that are involved in user data transmission in a top-down approach. Protocols that are used on top like TCP/IP or used as link protocols in the fixed network will not be introduced, since extensive literature already exists in that field.

3.1 GPRS Tunneling Protocol (GTP)

In GPRS networks the interworking with TCP/IP or X.25 based networks is foreseen [10] on the G_i interface. Between SGSN and GGSN on the G_n interface, encapsulated data packets are transmitted with the help of the GTP. GTP signaling is used by the *GPRS Mobility Management* (GMM)/*Session Management* (SM) (see Chapter 4) to

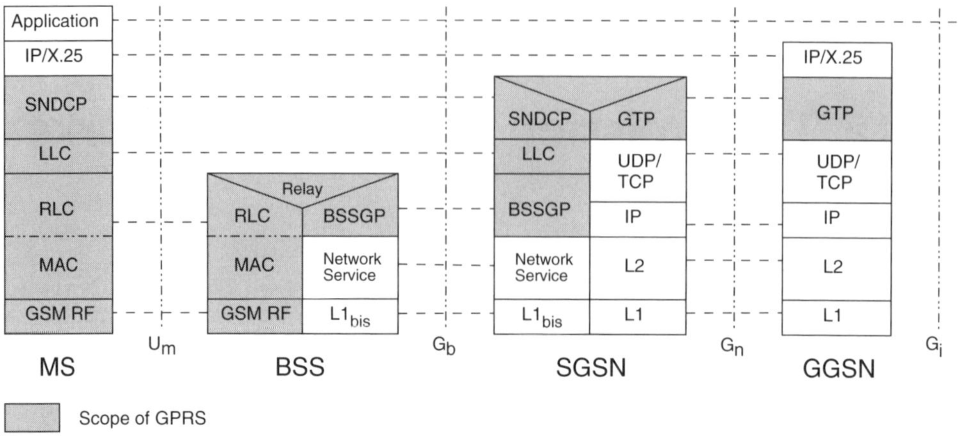

Figure 3.1: GPRS user plane

create, modify and delete tunnels [9]. GTP uses the *User Datagram Protocol* (UDP) for the transmission in the core network.

Tunnels are used to carry encapsulated PDUs between a given GSN pair for individual mobile stations (see Figure 3.2). The *Tunnel Identifier* (TID), part of the GTP header, indicates to which tunnel a particular datagram belongs. The TID specifies the PDP context (see Chapter 4) and is composed of the *Mobile Country Code* (MCC), *Mobile Network Code* (MNC) and parts of the *International Mobile Subscriber Identity* (IMSI) (see Section 2.1.5) and *Network layer Service Access Point Identifier* (N-SAPI). Like this, datagrams injected into the GPRS network at the GGSN are tunneled to the correct SGSN area, where the MS is presently located, which means that mobility-related routing is completely supported by the GTP. In other mobile core networks, e.g., networks based on the 3GPP2 specifications, this function is realized by *Mobile IP* [94].

The GTP comprises two different entities the GTP-U entity that is responsible for user data tunneling and the GTP-C entity responsible for signaling between SGSN and GGSN (see Section 4.3 and Section 4.2.3.4) using path, tunnel, location and mobility management messages.

3.2 Base Station Subsystem GPRS Protocol (BSSGP)

On the G_b interface, the BSSGP [15] provides a connectionless link between BSS and SGSN. This protocol's main task is the flow control for the downlink transfer of LLC PDUs. No flow control is performed in the uplink direction. The SGSN is expected to accept all data that is sent to it—the buffers and link capacity have to be dimensioned according to this premise to avoid loss of uplink data.

3.2.1 Flow Control between SGSN and BSS

Each cell in the coverage area of one BSS is virtually connected to the SGSN by a *BSSGP Virtual Connection* (BVC), identified by a *BSSGP Virtual Connection Identifier* (BVCI). The BSS maintains one queue for each BVCI that contains the

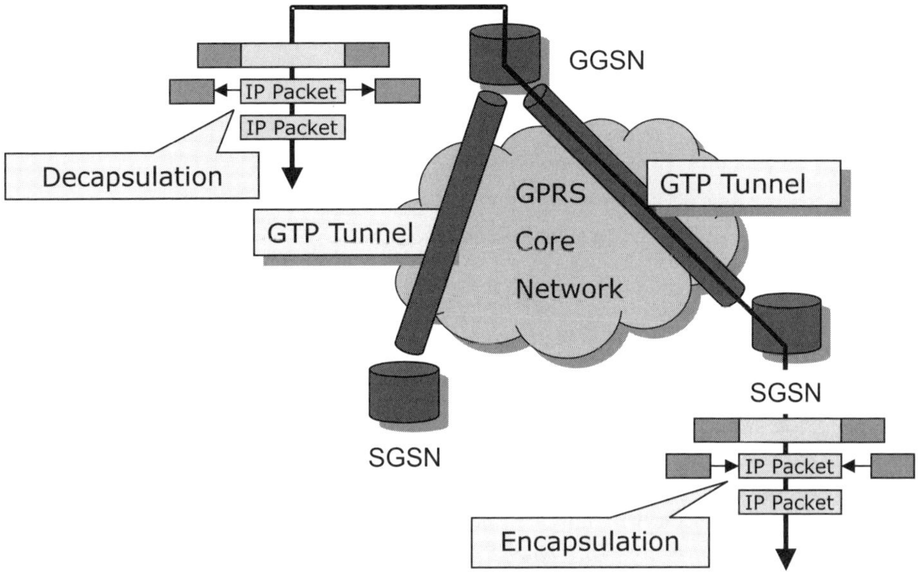

Figure 3.2: IP tunnel in the GPRS core network

LLC PDUs for that particular cell. There are several ways to further split these queues, e.g., into one queue per MS, or into one queue that serves all LLC PDUs belonging to a certain delay or precedence class. The queues are filled with downlink LLC PDUs and emptied by the BSS forwarding these PDUs to the MSs addressed. The SGSN is regularly informed by the BSS about the maximum queue size available for each BVC and MS at a given time, and also about the rate at which the available queue size increases. Thus, the SGSN is enabled to estimate the maximum allowable throughput per BVC as well as per MS belonging to the BVC.

Within the SGSN there are queues provided, similar to those in the BSS. Here the downlink LLC PDUs are queued and scheduled as long as the maximum allowable throughput per BVC and per MS is not exceeded. By this mechanism, it is ensured that both the MS and the BSS are able to handle the incoming traffic.

3.2.2 BSS Context

The SGSN can provide a BSS with information related to ongoing user data transmission. The information related to one MS is stored in a BSS context. The BSS may contain BSS contexts for several MSs. A BSS context contains a number of BSS *Packet Flow Contexts* (PFCs). Each BSS PFC is identified by a *Packet Flow Identifier* (PFI) assigned by the SGSN. A BSS PFC is shared by one or more activated *Packet Data Protocol* (PDP) contexts with identical or similar negotiated QoS profiles. The data transmission related to PDP contexts that share the same BSS PFC comprises one packet flow.

Three packet flows are predefined, and identified by three reserved PFI values. The BSS does not negotiate BSS PFCs for these pre-defined packet flows with the SGSN. One pre-defined packet flow is used for best-effort service, one is used for SMS, and

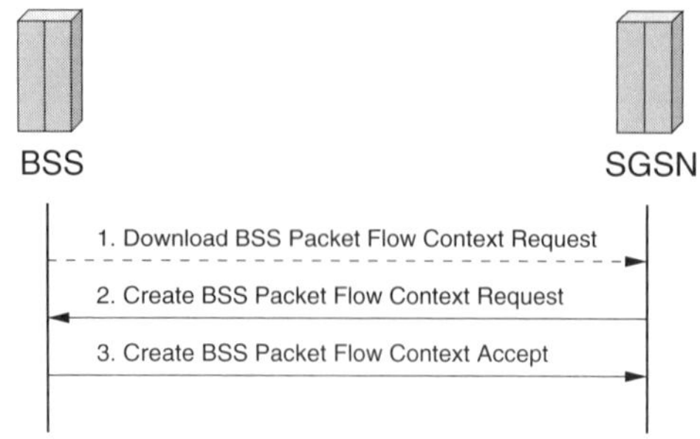

Figure 3.3: BSS *Packet Flow Context* (PFC) creation procedure

one is used for signaling. The SGSN can assign the best-effort or SMS PFI to any PDP context. In the SMS case, the BSS handles the packet flow for the PDP context with the same QoS that it handles SMS with.

The combined BSS QoS profile for the PDP contexts that share the same packet flow is called the *Aggregate BSS QoS Profile* (ABQP). The ABQP is considered to be a single parameter with multiple data transfer attributes. It defines the QoS that must be provided by the BSS for a given packet flow between the MS and the SGSN, i.e., for the U_m and G_b interfaces combined. The ABQP is negotiated between the SGSN and the BSS.

A BSS packet flow timer indicates the maximum time for which the BSS may store the BSS PFC. The BSS packet flow timer is started when the BSS PFC is stored in the BSS and when an LLC frame is received from the MS. When the BSS packet flow timer expires the BSS deletes the BSS PFC.

When a PDP context is activated, modified, or deactivated (see Section 4.3.4), the SGSN may create, modify, or delete BSS PFCs.

3.2.3 BSS Packet Flow Context Creation

On receiving a request to transmit an uplink or downlink LLC PDU for which no BSS PFC exists in the BSS, the BSS may request the download of the BSS PFC from the SGSN.

The SGSN may at any time request the creation of a BSS PFC, e.g. due to the activation of a PDP context.

The BSS PFC creation procedure is illustrated in Figure 3.3. If the BSS receives a request to transfer an uplink or downlink user data LLC PDU without having BSS PFC available for this flow, it sends a `Download BSS Packet Flow Context Request` message to the SGSN. Until the BSS receives the BSS PFC, the BSS will handle uplink and downlink transfers according to a default ABQP. For uplink transfers, the default profile is specific to the radio priority level.

On reception of the `Download BSS Packet Flow Context Request`, or in PDP context activation or modification procedures, the SGSN sends a `Create BSS Packet Flow Context Request` message to the associated BSS.

The BSS may restrict the requested ABQP given its capabilities and the current load. It then creates a BSS PFC, inserts the parameters in its BSS context, and returns a `Create BSS Packet Flow Context Accept` message to the SGSN. The BSS uses

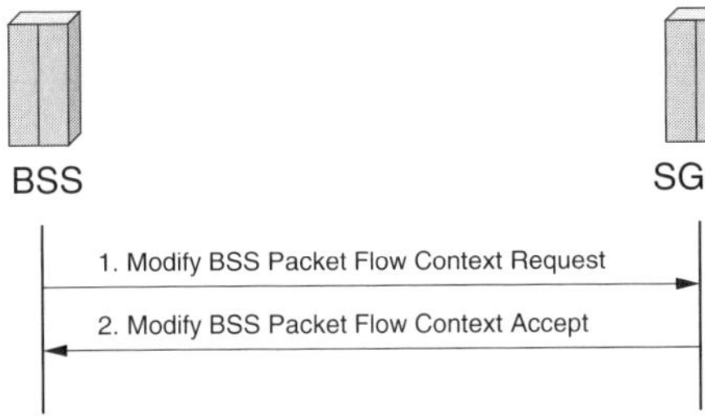

Figure 3.4: BSS-initiated BSS PFC modification procedure

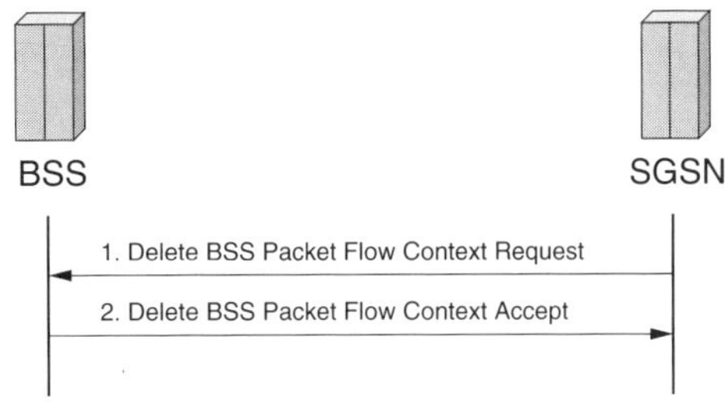

Figure 3.5: SGSN-initiated BSS PFC deletion procedure

the negotiated ABQP when allocating radio resources and other resources such as buffer capacity.

3.2.4 BSS Packet Flow Context Modification

If the SGSN wishes to modify the contents of any existing BSS PFC, e.g. due to the activation, modification, or deactivation of a PDP context, it uses the BSS PFC creation procedure. The BSS will instead of creating a BSS PFC overwrite the existing parameters with the modified parameters.

The BSS itself can at any time request modification of the contents of an existing BSS PFC, e.g. due to a change in the resource availability at the BSS. Figure 3.4 illustrates the relevant procedure. The SGSN may again restrict the requested ABQP.

3.2.5 BSS Packet Flow Context Deletion

The BSS can, owing to, e.g. memory restrictions, at any time delete a BSS PFC without notifying the SGSN.

The SGSN may request the deletion of a BSS PFC with the SGSN-initiated BSS PFC deletion procedure, as illustrated in Figure 3.5.

3.3 Sub-Network Dependent Convergence Protocol (SNDCP)

Network layer protocols are intended to be capable of operating over services derived from a wide variety of subnetworks and data links. GPRS supports several network layer protocols providing protocol transparency for the users of the service. Therefore, all functions related to transfer of network layer PDUs are carried out in a transparent way by the GPRS network entities. SNDCP provides this adaptation of different network layers to the logical link by using N-SAPIs to map different PDPs onto the services provided by the LLC layer. Additionally SNDCP supports compression of redundant user data and protocol control information. The main functions of the SNDCP layer are [8]:

- Multiplexing of several PDPs.
- N-PDU buffering.

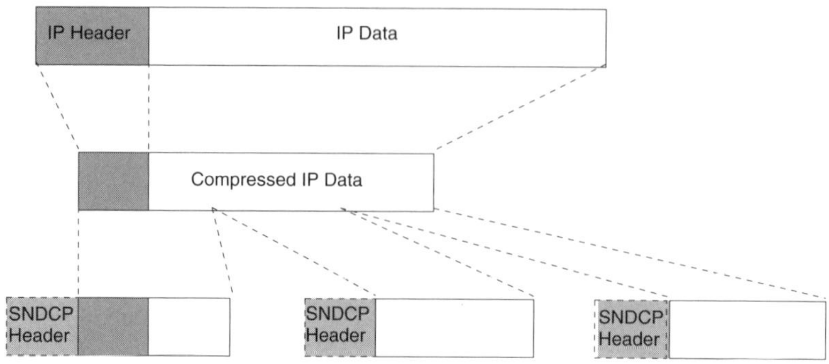

Figure 3.6: SNDCP data flow

- Compression and decompression of user data and control information (e.g., TCP/IP header compression).
- Segmentation and reassembly of network PDUs (see Figure 3.6).

The N-PDUs are buffered in the SNDCP layer before they are compressed, segmented and transmitted to the LLC layer. For acknowledged data transfer, the SNDCP entity buffers an N-PDU until successful reception of all segments of the N-PDU have been confirmed. The confirmation is carried out by the underlying layer using a confirm service primitive. Buffered N-PDUs which have been completely received are discarded. For unacknowledged data transfer, the SNDCP deletes an N-PDU immediately after it has been delivered to the LLC layer. Acknowledged or unacknowledged mode on SNDCP level only means that the SNDCP layer is using the acknowledged or unacknowledged data service from the LLC layer. In the SNDCP layer itself no error handling functions are performed.

The protocol control information compression method is specific for each network layer protocol type. TCP/IP (IPv4) header compression is specified in [78]. For data compression [70] may be used.

After the SNDCP header is added the compressed N-PDU is segmented if it is longer than the maximum payload size for an LLC frame.

3.4 Logical Link Control (LLC)

The LLC layer [7] provides a highly reliable and ciphered link between the MS and the SGSN. It includes functions for:

- Provision of one or more logical connections.
- Sequence control.
- Error detection and correction.
- Flow control.
- Ciphering.

The protocol for the LLC sublayer is based on well-known data link protocols like the *Link Access Procedure on the D-channel* ($LAPD_m$) used in GSM and the *High level Data Link Control* (HDLC) protocol standardized by ISO for *Open Systems Interconnection* (OSI).

The key modifications can be summarized as follows:

Variable frame length The GPRS protocol architecture allows a variable frame length at the LLC level. Therefore frame delimiters and bit stuffing are not necessary but an additional field is required in the frame header for specification of the frame length.

Prioritized Service Access Points (SAPs) The priority classes in GPRS are considered in the introduction of new SAPs with priorities.

3.4.1 Layer Entities and Service Access Points

The LLC layer provides six SAPs to the upper layer:

- Four SAPs are dedicated to the SNDCP that manages data packet transmission; one SAP exists for each QoS.
- One SAP is dedicated to *GPRS Mobility Management* (GMM) (see Chapter 4)
- One SAP is dedicated to *Short Message Service* (SMS)

The related services are performed by different logical link entities (see Figure 3.7) that are managed by the logical link managment entity. Corresponding to four radio priority levels, four *Service Access Point Identifiers* (SAPIs) are required for packet data transmission. SAPIs of the LLC layer identify a point at which LLC services are provided by an LLC entity to a layer-3 entity. With the introduction of service or subscriber differentiation, it is also necessary for flow control and error correction to be controlled on a priority basis.

3.4.2 Logical Link Identifiers

Two different logical link identifiers exist in the GPRS protocol architecture, the *Temporary Logical Link Identifier* (TLLI) and the *Data Link Connection Identifier* (DLCI). When the MS attaches to the GPRS network, a TLLI is assigned to the MS. A one-to-one relationship exists between the TLLI and the *International Mobile Subscriber Identity* (IMSI) (called *Packet Temporary Mobile Subscriber Identity* (P-TMSI) in GPRS, see Section 2.1.5) that is only known in the MS and the SGSN. The DLCI is the identifier of the LLC connection and is composed of the TLLI and the LLC SAPI.

3.4.3 LLC Frame Structure

The LLC frame format is shown in Figure 3.8. The frame header consists of the address field and the control field and ranges from two to 37 octets. The address field consists of a single octet containing the SAPI (see Section 3.4.1) and the identifier of the logical link, the DLCI. The control field typically consists of between one and three octets (see Figure 3.9). LLC frames can be classified into information

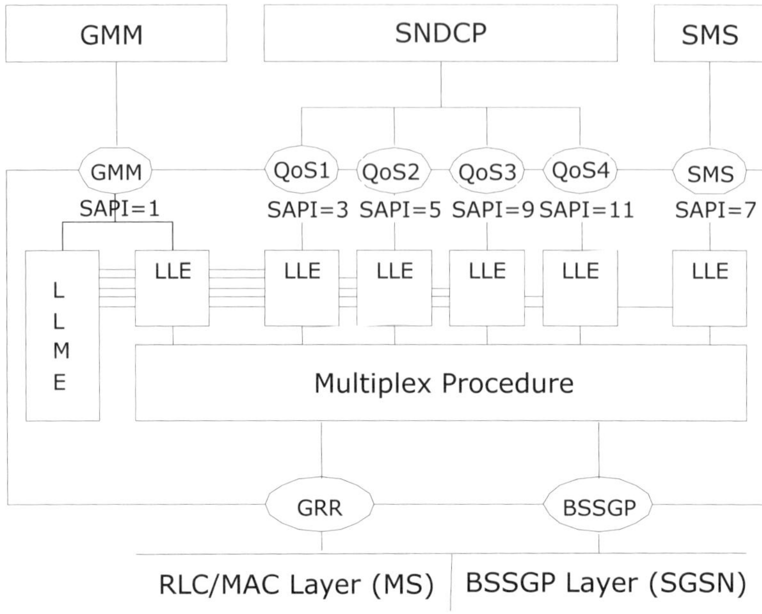

Figure 3.7: LLC layer structure

frames (I frames) containing sequence numbers for acknowledged data transmission, unnumbered information frames (UI frames) for unacknowledged data transmission and supervisory frames (S frames) and unnumbered frames (U frames) containing LLC control messages. The *Selective Acknowledgement* (SACK) supervisory frame additionally includes a bitmap field of variable length of up to 32 octets. The information field, if present, follows the control field. The maximum length of the information field depends on the SAPI. The *Frame Check Sequence* (FCS) field consists of a 24-bit *Cyclic Redundancy Check* (CRC) code that is used to detect bit errors in the frame header and information field.

3.4.4 LLC Frame Transmission

The LLC layer performs an ARQ protocol based on retransmissions after timeouts or frame loss detections and optionally uses bitmap-based selective acknowledgements.

3.4.4.1 LLC Modes

LLC can operate in acknowledged and unacknowledged mode. In unacknowledged mode the network layer PDUs are transmitted in *Asynchronous Disconnected Mode* (ADM) using UI command frames (see Figure 3.10). Neither LLC error recovery nor reordering procedures are defined, but transmission and format errors are detected. Duplicate UI frames are discarded. Flow control procedures are not defined in unacknowledged mode. In acknowledged mode network layer information is transmitted in order in numbered I frames that are acknowledged by the peer entity. Error recovery and reordering procedures based on retransmission of unacknowledged

8 7 6 5 4 3 2 1
Address Field (1 octet)
Control Field (variable length, max. 36 octets)
Information Field (variable length, max. N201 octets)
Frame Check Sequence Field (3 octets)

Figure 3.8: LLC frame format

Control Field Bits

<table>
<tr><th></th><th>8</th><th>7</th><th>6</th><th>5</th><th>4</th><th>3</th><th>2</th><th>1</th><th>Octet</th></tr>
<tr><td rowspan="3">I format (I+S)</td><td>0</td><td>A</td><td>X</td><td colspan="5">N(S)</td><td>1</td></tr>
<tr><td colspan="4">N(S)</td><td>X</td><td colspan="3">N(R)</td><td>2</td></tr>
<tr><td colspan="6">N(R)</td><td>S_1</td><td>S_2</td><td>3</td></tr>
<tr><td rowspan="2">S format</td><td>1</td><td>0</td><td>A</td><td>X</td><td>X</td><td colspan="3">N(R)</td><td>1</td></tr>
<tr><td colspan="6">N(R)</td><td>S_1</td><td>S_2</td><td>2</td></tr>
<tr><td rowspan="2">UI format</td><td>1</td><td>1</td><td>0</td><td>X</td><td>X</td><td colspan="3">N(U)</td><td>1</td></tr>
<tr><td colspan="6">N(U)</td><td>E</td><td>PM</td><td>2</td></tr>
<tr><td>U format</td><td>1</td><td>1</td><td>1</td><td>P/F</td><td>M_4</td><td>M_3</td><td>M_2</td><td>M_1</td><td>1</td></tr>
</table>

A	Acknowledgement request bit
E	Encryption function bit
M	Unnumbered function bit
N(R)	Transmitter receive sequence number
N(S)	Transmitter send sequence number
N(U)	Transmitter unconfirmed sequence number
P/F	Poll bit, when issued as a command Final bit, when issued as a response
PM	Protected mode bit
S	Supervisory function bit
X	Spare bit

Figure 3.9: LLC control field

I frames are specified. Acknowledged operation requires that the *Asynchronous Balanced Mode* (ABM) has been initiated in an establishment procedure using the *Set Asynchronous Balance Mode* (SABM) command (see Figure 3.11).

3.4.4.2 Flow Control

The flow control between the LLC peer entities is determined by the LLC window size that ranges from 2 to 16 and a modulus of 64 frames.

To perform vertical flow control between LLC and RLC/MAC, i.e. not to overload the RLC/MAC layer with I-frames, a maximum buffer size is introduced. The actual buffer size is incremented with the length of the information field of a transmitted I frame and decremented with the size of an acknowledged LLC frame. The value of the buffer size shall never exceed the maximum. If a new I-frame to transmit would make the buffer size exceed the maximum, then the LLC entity shall not transmit any new I frames, but still may retransmit I frames as a result of the error recovery procedures.

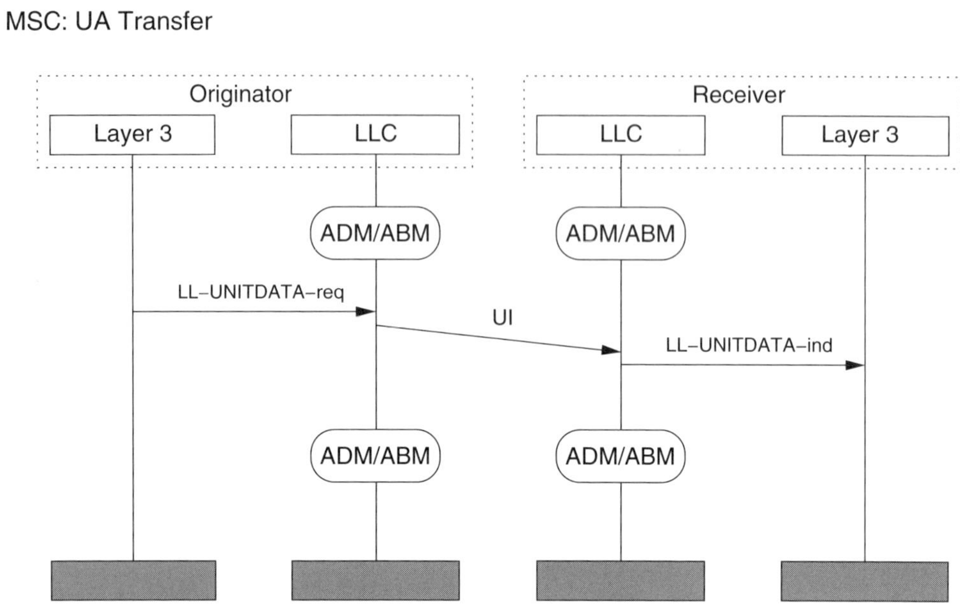

Figure 3.10: LLC unacknowledged mode

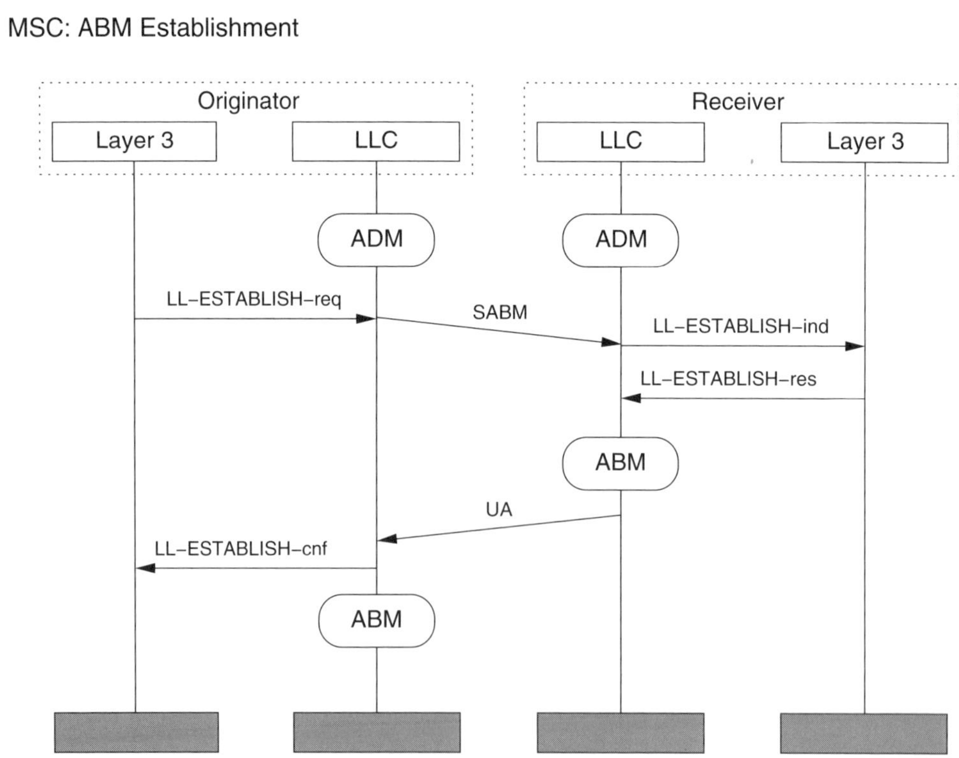

Figure 3.11: LLC acknowledged mode

3.4.5 Ciphering

In GPRS, the ciphering is no longer performed between the BTS and the MS as in GSM, but at the LLC level between SGSN and the MS. As in GSM a ciphering key is produced with the ciphering algorithm from the identification key and a non-predictable random number generated by the *Authentication Center* (AuC) (see Section 2.1.3).

3.5 Radio Link Control (RLC) and Medium Access Control (MAC)

The idea of packet-switched services in GSM is to multiplex several users onto one single physical channel in order to use its capacity in a more efficient way. Furthermore, one single MS can use more than one PDCH simultaneously to increase the data rate. The maximum number of PDCHs that can be used in parallel is determined by the multislot capabilities of the MS, which can range in uplink and downlink from multislot capability 1 (one PDCH) up to multislot capability 8 (all eight PDCHs). The *multislot class* specified in the GPRS standard is a combination of the multislot capabilities in the uplink and downlink.

The *Data Link Control* (DLC) layer at the mobile U_m interface is divided in two sublayers: the RLC layer and the MAC layer, see Figure 3.1. The RLC sublayer provides a reliable logical connection between MS and BSS, while the MAC sublayer controls the access to the physical medium, the radio link.

3.5.1 Multiplexing Principles

The radio interface consists of asymmetric and independent uplink and downlink channels. The downlink carries packet data from the network to multiple MSs and does not require contention arbitration. The uplink is shared among multiple MSs and requires contention control procedures.

The access to the uplink is realized with a *Slotted-Aloha* based reservation protocol. A `Packet Channel Request` sent by an MS on the *Packet Random Access Channel* (PRACH) is responded by a `Packet Uplink Assignment` message indicating the uplink resources reserved for the MS.

LLC frames are segmented into RLC data blocks. At the RLC/MAC layer, a selective ARQ protocol between the MS and the network provides retransmission of erroneous RLC data blocks. When a complete LLC frame is successfully transferred across the RLC layer, it is forwarded to the peer LLC entity.

A *Temporary Block Flow* (TBF) is a virtual connection to support the unidirectional transfer of LLC PDUs on the packet data physical channels. The TBF is the allocated radio resource on one or more PDCHs and comprises a number of RLC/MAC blocks carrying one or more LLC PDUs. A TBF is temporary and maintained only for the duration of the data transfer, i.e., until there are no more RLC/MAC blocks to be transmitted and, in RLC acknowledged mode, all of the transmitted RLC/MAC blocks have been successfully acknowledged by the receiving entity.

A *Temporary Flow Identity* (TFI) is assigned for each TBF by the network. The MS assumes that the TFI value is unique among the concurrent TBFs in each direction. The same TFI value may be used concurrently for TBFs in opposite directions. An

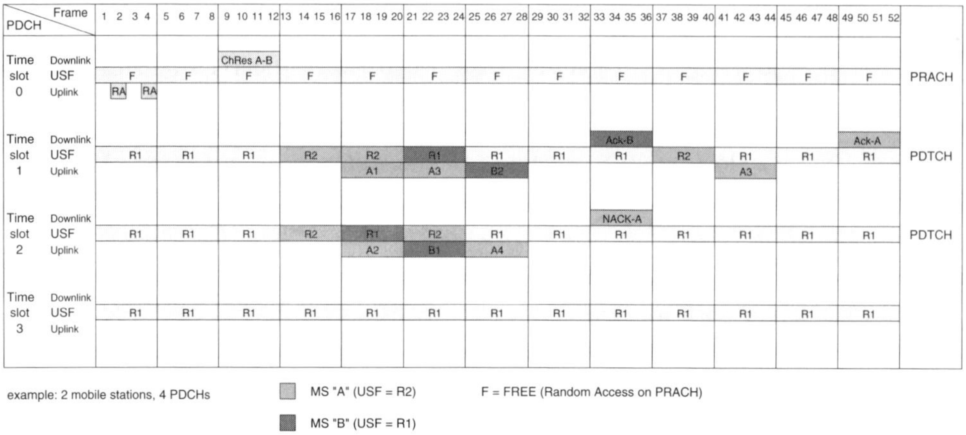

Figure 3.12: Channel map for uplink packet data traffic

RLC/MAC block associated with a certain TBF comprises the TFI. The TBF is identified by the TFI together with the direction in which the RLC data block is sent and in case of an RLC/MAC control message additionally with the message type.

An *Uplink State Flag* (USF) is included in the header of each RLC/MAC block on a downlink PDCH. The USF may be used on the PDCH to allow multiplexing of radio blocks from different MSs and comprises three bits at the beginning of each radio block that is sent on the downlink. It enables the coding of *eight* different USF states which are used to multiplex the uplink traffic. On PCCCH, one USF value is used to denote PRACH (USF = FREE). The *seven* other USF values are used to reserve the uplink for different MSs.

Three medium access modes are supported:

- Dynamic Allocation (characterized by a USF)
- Extended Dynamic Allocation (supports the multislot functionality)
- Fixed Allocation

The Extended Dynamic Allocation medium access method extends the Dynamic Allocation medium access method to allow higher uplink throughput. The network allocates to the MS a subset of 1 to N consecutive PDCHs, where N depends on the multislot class of the MS.

In `packet idle mode` the MS monitors the relevant paging subchannels on the PCCCH. In `packet idle mode` no TBF exists and the upper layer may require the transfer of an LLC PDU, which implicitly triggers the establishment of a TBF and the transition to the `packet transfer mode`.

In `packet transfer mode` a TBF provides a physical point-to-point connection on one or more packet data physical channels for the unidirectional transfer of LLC PDUs between the network and the MS. Continuous transfer of one or more LLC PDUs is possible. Two parallel TBFs may be established in opposite directions for each MS.

An uplink transfer using extended dynamic allocation with two downlink slots per MS is shown in Figure 3.12 to illustrate the GPRS multiplexing principle. After the

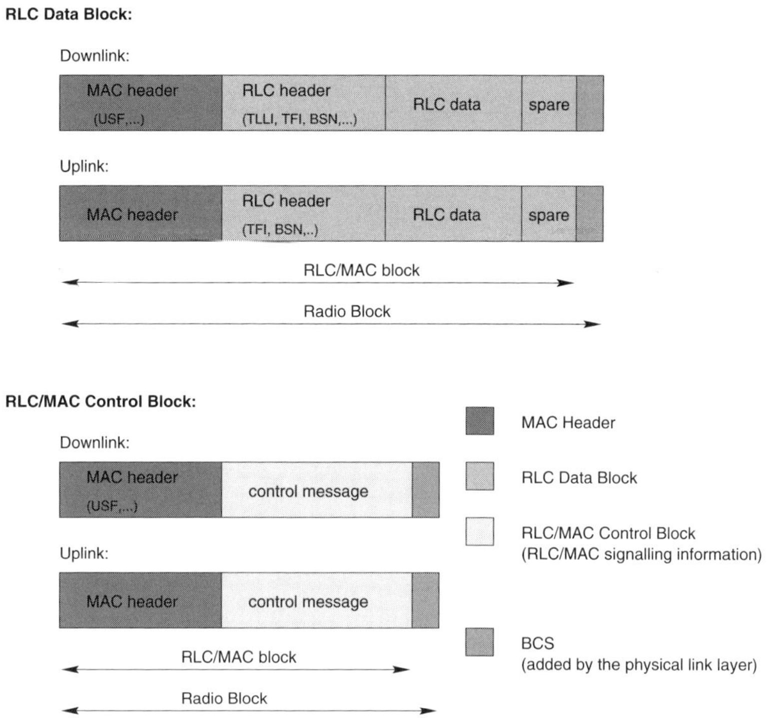

Figure 3.13: Structure of the radio block and RLC/MAC block

Random Access (RA) of the two MSs A and B the PDCHs 1 and 2 are allocated for both MSs using a Channel Reservation (CR) message. The USFs R2 and R1 are allocated to MS A and B, respectively. After the indication of the related USF on the downlink the MSs can transmit uplink radio blocks in the following uplink radio block periods. The radio blocks are acknowledged by the BS.

3.5.2 RLC/MAC Block Structure

An RLC/MAC block consists of a MAC Header and an RLC data block or RLC/MAC control block, respectively. In the PLL a *Block Check Sequence* (BCS) is added for error detection. The RLC/MAC block together with the BCS are building one radio block that is transmitted in one PDCH in one radio block period (see Section 2.2.5.1). The RLC/MAC block structure is shown in Figure 3.13.

The RLC data block consists of RLC header, an RLC data field and spare bits. Each RLC data block may be encoded using any of the available channel coding schemes CS-1, CS-2, CS-3 or CS-4. The size of the RLC data block for each of the channel coding schemes is shown in Table 3.1. Two octets are always used in the RLC header for TFI, *Final Block Indicator* (FBI) and *Block Sequence Number* (BSN). The field for the `Length Indicator` and the `Extension`-Bit needs at least one octet. One additional octet is needed for each additional LLC frame. The remaining space is used for user data.

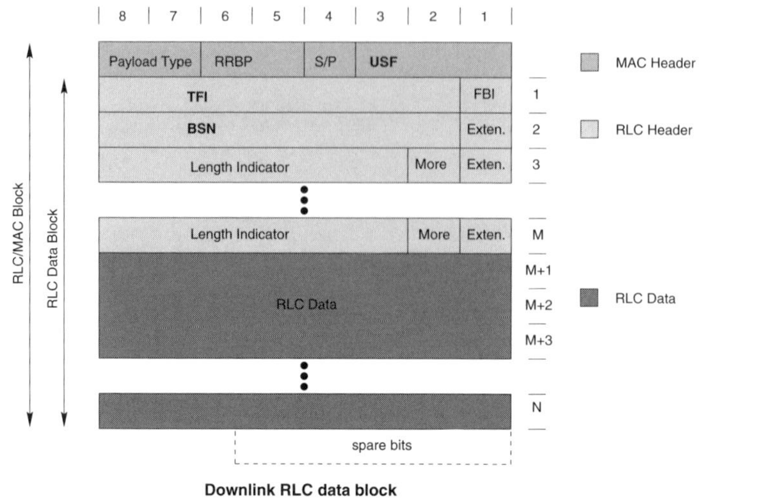

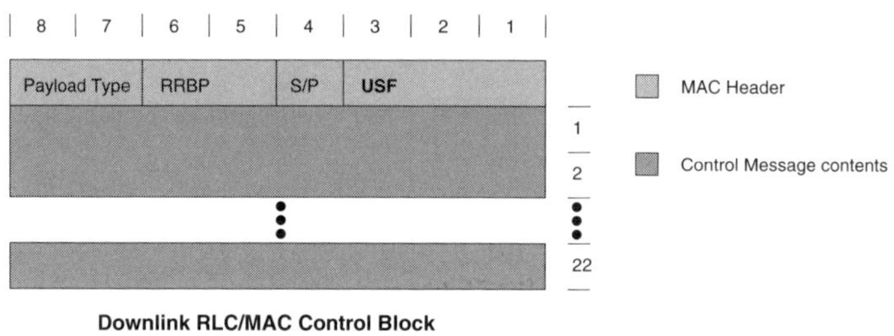

Figure 3.14: Block structure in downlink

Table 3.1: RLC block size

Coding Scheme	RLC info field size [byte]	RLC data block size (without spare bits) [byte (bit)]	Spare bits [bit]	RLC/MAC block size [bit]
CS-1	20	22 (176)	0	181
CS-2	30	32 (256)	7	268
CS-3	36	38 (304)	3	312
CS-4	50	52 (416)	7	428

The RLC/MAC blocks containing an RLC/MAC control block always consist of 23 bytes, one byte for the MAC header and 22 bytes for the RLC/MAC control block containing the control message.

Each radio block carrying an RLC data block can be coded with one of the four different Channel Coding Schemes CS-1 to CS-4. For a radio block containing an RLC/MAC control block always the coding scheme CS-1 is used. The exception is messages that use the existing access burst, e.g., the `Packet Channel Request` message. The four different coding schemes are explained in detail in Section 3.5.7.

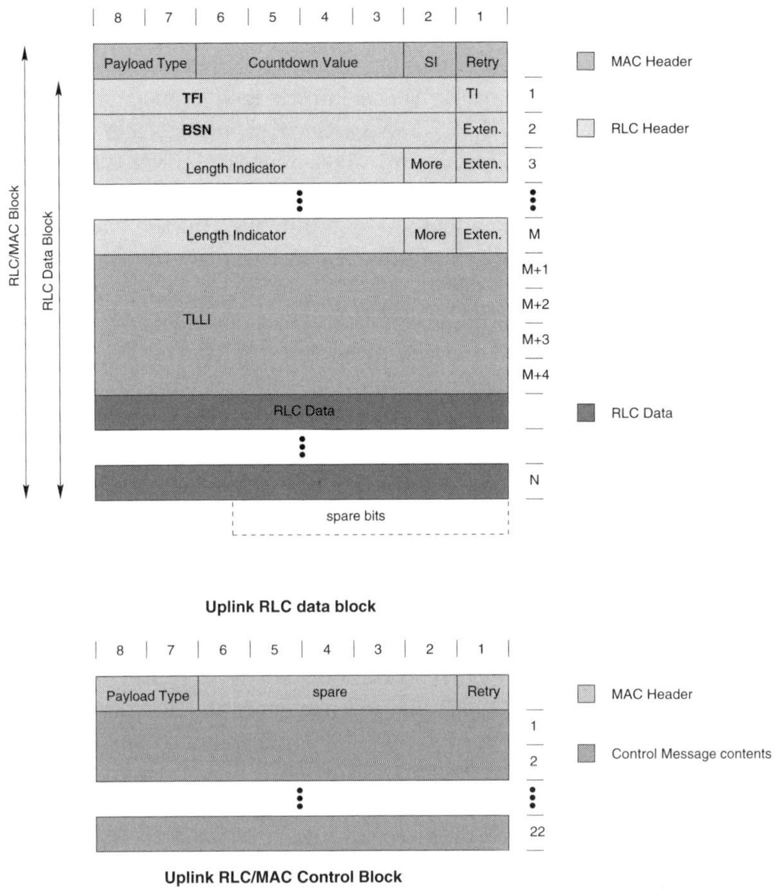

Figure 3.15: Block structure in uplink

3.5.3 RLC Functions

The RLC functions in GPRS provide an interface towards the LLC layer, especially the segmentation and reassembly of LLC PDUs into RLC data blocks depending on the used *Coding Scheme* (CS) (see Table 3.1). *Backward Error Correction* (BEC) is used to enable the selective retransmission of uncorrectable PDUs. The BCS for error detection in GPRS is provided already by the Physical Link Layer. There are two possible modes provided by the RLC/MAC layer: the acknowledged mode, used for reliable data transmission, and the unacknowledged mode, mainly used for real-time services, such as video or voice, where time delay is most critical, whereas bit errors are less important unless they pass the accepted range. For RLC acknowledged mode, a selective ARQ protocol is used between the MS and the BS, which provides retransmission of erroneous RLC data blocks. As soon as a complete LLC frame is successfully transferred across the RLC layer, it is forwarded to the peer LLC entity.

3.5.3.1 Data Transmission in RLC Acknowledged Mode

In this mode, the transmitting side numbers the RLC data blocks with the *Block Sequence Number* (BSN) to supervise the acknowledged transmission of the RLC blocks. With *Acknowledgement* (ACK)/*Negative Acknowledgement* (NACK) messages on uplink and downlink (`Packet Uplink/Downlink ACK/NACK`), the receiving side can request retransmissions of RLC blocks if needed.

Since the radio resource management is located in the BS, the transmission of the acknowledgement messages in both directions is controlled by the BSS, since the BSS manages the resource in the radio cell. The MS sends the ACK/NACK message in the reserved radio block which is requested from the peer entity by polling. `Packet Uplink ACK/NACK` messages can be transmitted directly by the BS on the PACCH. In case of a negative acknowledgement, only those blocks listed as erroneous are retransmitted. In the case that the TBF ends before the LLC frame is transmitted completely, the missing part has to be transmitted during the new TBF.

Release of the uplink TBF The mobile station sends the *Countdown Value* (CV) in each uplink RLC data block to indicate to the network the absolute BSN of the last RLC data block that will be sent in the uplink TBF. The network is then able to calculate the number of remaining RLC data blocks for the current uplink TBF. The final RLC data block transmitted in the TBF has CV set to the value 0.

If the last RLC block is sent with the CV value set to 0 and if there are no more unacknowledged RLC blocks to transmit, the timer `T3182` is started. This timer is on the reception of a `Packet Uplink ACK/NACK` message. If this `Packet Uplink ACK/NACK` message orders more retransmissions, the MS has to wait until the uplink resources are reallocated before transmitting the requested RLC data blocks. The network sets the Final Ack Indicator when all data has been received properly and causes the MS to release the TBF. The network also includes a valid *Relative Reserved Block Period* (RRBP) field in the RLC block header and waits for the `Packet Channel Request`, which is specified by the RRBP. This RRBP value specifies a single uplink block in which the MS transmits a `Packet Channel Request` message or an associated control message (e.g., ACK) to the network.

Release of the downlink TBF The release of the downlink TBF is initiated by the BS by setting the FBI to the value 1 and sending the RLC block with a valid RRBP field. Additionally, the timer `T3191` is started to define when the current assignment is invalid on the mobile side so that the TFI can be reused.

The network is now waiting for the `Packet Downlink ACK/NACK` message, which indicates whether some retransmissions are required or not. If there are any, the network begins to retransmit them according to the ARQ protocol. If no retransmissions are necessary, the TBF is released.

3.5.3.2 Data Transmission in RLC Unacknowledged Mode

This mode comprises no retransmissions. The BSN is now only used to reassemble the RLC blocks. The `Packet ACK/NACK` messages are sent to convey the necessary control signaling, as for instance the monitoring of the channel quality. The MS transmits RLC data blocks without receiving a `Packet ACK/NACK` message until the *Window Size* (WS) is reached. Then it starts the timer `T3182` to wait for a `Packet Uplink`

ACK/NACK message. On receipt of this message, the timer will be stopped. On expiry of the timer, an abnormal release with cell reselection will be performed.

Release of the uplink TBF The release of the uplink TBF is initiated by the MS, when it sends the final RLC block with CV set to 0. The network on the other side starts timer T3169 and sends a Packet Uplink ACK/NACK message with the Final Acknowledge Indicator set to 1.

Release of the downlink TBF Initiated by the network, the release of the downlink TBF is indicated with the enabled FBI field in the header of a downlink RLC block and with a valid RRBP field. Timer T3191 is also started in the network. Receiving this RLC block, the MS transmits the Packet Channel Request message in the uplink block indicated by the RRBP, starts timer T3192 and continues to monitor all assigned PDCHs. When receiving a subsequent RLC block with a valid RRBP and the FBI set to 1, the MS retransmits the Packet Downlink ACK/NACK. On expiry of T3191, the network terminates the TBF and releases all resources. On expiry of timer T3192, the MS stops monitoring the assigned downlink PDCH and monitors the AGCH and PCH, or the PAGCH and PPCH.

Abnormal release In an abnormal release case, the sending side begins to establish a new RLC connection with a new TFI. At the beginning of the new TBF, the old RLC blocks, which could not be sent within the old TBF, will be transmitted.

3.5.3.3 ARQ Procedure

The transfer of RLC data blocks in the RLC acknowledged mode is supported by the retransmission of erroneous RLC data blocks. The transmitting side numbers the RLC data blocks via the BSN for retransmission and for reassembly. Each RLC data block contains a BSN.

Transmit window The *send state variable* $V(S)$ denotes the block sequence number (BSN) of the next in-sequence RLC data block to be transmitted. At the time that an in-sequence RLC data block is determined for transmission, the value of BSN is set equal to the value of the send state variable $V(S)$. The value of $V(S)$ is incremented by 1 after transmission of the RLC data block with BSN $= V(S)$.

The *acknowledge state variable* $V(A)$ contains the BSN value of the oldest RLC data block that has not been positively acknowledged by its peer. The value of $V(A)$ is updated from the value received from its peer in the *Received Block Bitmap* (RBB) of the Packet Ack/Nack (Packet Uplink Ack/Nack or Packet Downlink Ack/Nack) message.

In *acknowledged RLC mode*, $V(S)$ shall not exceed $V(A)$ modulo 128 by more than the window size k. This means:

$$[\ V(S) - V(A)\ ]\ modulo\ 128 \leq k$$

In acknowledged RLC mode all BSNs which meet the following criteria are valid within the transmit window:

$$[\ V(A) \leq BSN < V(S)\ ]\ modulo\ 128$$

The inequality defines the transmit window in acknowledged mode; in *unacknowledged mode*, all values of BSN are within the transmit window (see Figure 3.16).

When an element in $V(B)$ falls outside the active transmit window, the element is set to the value `INVALID`.

The *acknowledge state array* $V(B)$ is an array of 128 elements indicating the acknowledgement status (`NACKED`, `PENDING ACK`, `ACKED`, `INVALID`) of k previous RLC data blocks. The array is indexed relative to the acknowledged state variable $V(A)$ modulo 128 or relative to the *Starting Sequence Number* (SSN). The values of $V(B)$ are updated from the values received from its peer in the RBB of the `Packet Ack/Nack` message.

The `Packet Ack/Nack` message contains a SSN and an RBB. The message is sent by the RLC receiver and is received by the RLC transmitter. The SSN and RBB do not have a meaning in the RLC unacknowledged mode; they will be ignored by the receiver.

In Figure 3.17 the sending order of the radio blocks stored in the transmit window is shown. If the transmit window is not stalled, the transmitter transmits the oldest RLC block whose acknowledgement status in $V(B)$ indexed relative to $V(A)$ has the value `NACKED` and sets the corresponding status in $V(B)$ of the transmitted RLC data block to the value `PENDING_ACK`. If no RLC data block with a `NACKED` status exists, the RLC data block with $BSN = V(S)$ is transmitted and the corresponding status in $V(B)$ is set to the value `PENDING_ACK`. BSN' is the BSN of most recently received RLC data block.

If the transmit window is stalled, the sending side transmits the oldest RLC data block, whose corresponding value in $V(B)$ has the status `PENDING_ACK`, then the next oldest RLC data block with such a status is transmitted, etc. If all RLC data blocks

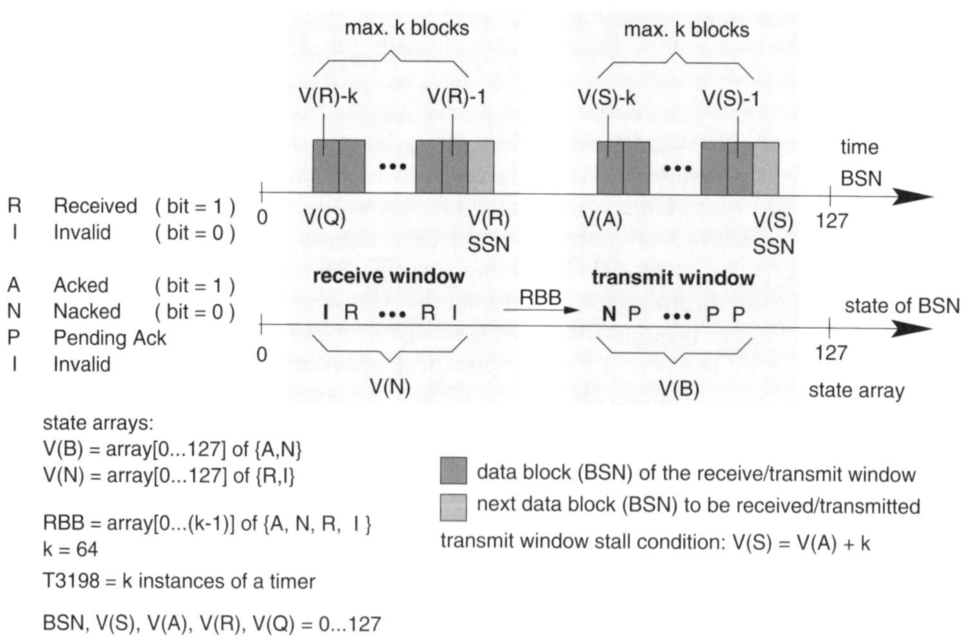

Figure 3.16: Receive window and transmit window

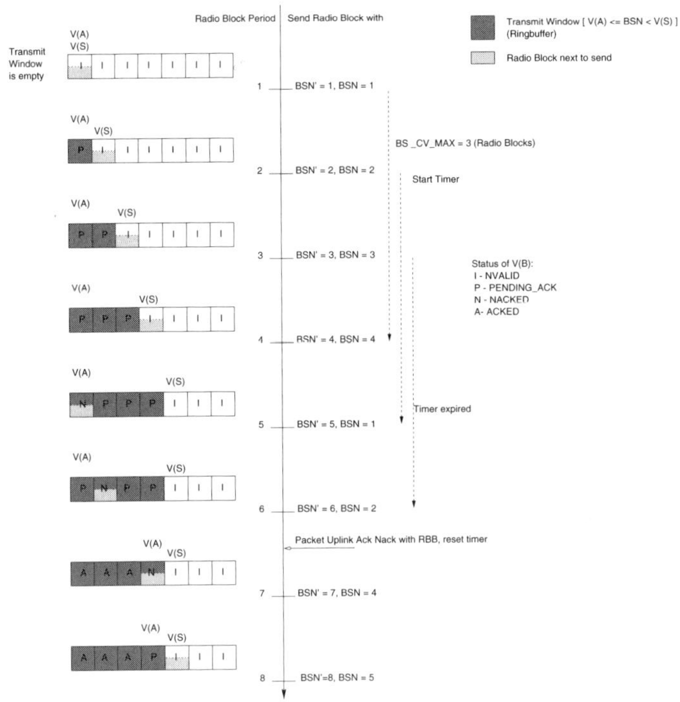

Figure 3.17: Sending order of the radio blocks in the transmit window

set to PENDING_ACK have been transmitted once, the process is repeated beginning with the oldest RLC data block. A maximum number of retransmissions for one radio block is not specified in the specification.

If the MS is the transmitter, an acknowledgement has to be received after BS_CV_MAX block periods for each RLC data block, otherwise the corresponding acknowledgement status is set to NACKED.

Receive window The *receive state variable* $V(R)$ denotes the BSN of the next in-sequence RLC data block expected to be received. The value of $V(R)$ is set to the greatest BSN value received in the current receive window, assuming that the RLC data block is error-free and the BSN does not exceed $[V(Q) + k]$ modulo 128.

The *receive window state variable* $V(Q)$ denotes the BSN of the oldest RLC data block within the receive window that has not been received. The value of $V(Q)$ is updated when the RLC receiver receives the RLC data block whose BSN is equal to $V(Q)$. The value of $V(Q)$ is then set to the value of the oldest BSN in the receive window that has not been received, or it is set to $V(R)$ if all RLC data blocks in the receive window have been received properly.

In *acknowledged RLC mode*, $V(R)$ shall not exceed $V(Q)$ modulo 128 by more than the window size k (Figure 3.16). This means:

$$[\ V(R) - V(Q)\ ]\ modulo\ 128 \leq k$$

In acknowledged RLC mode all BSNs which meet the following criteria are valid within the receive window:

$$[\ V(Q) \leq BSN < V(R)\ ]\ modulo\ 128$$

The inequality defines the receive window in acknowledged mode; in *unacknowledged mode*, all values of BSN are within the receive window.

The *receive state array* $V(N)$ is an array of 128 elements indicating the receive status (RECEIVED, INVALID) of k previous RLC data blocks. The array is indexed relative to the receive state variable $V(R)$ modulo 128. When an RLC data block is received with BSN valid in the receive window the corresponding element in $V(N)$ is set to the value RECEIVED. If an element in $V(N)$ falls outside the active window, the element is set to the value INVALID. This condition can be described by:

$$BSN\ modulo\ 128 < [\ V(R) - k\ ]\ modulo\ 128$$

Update of the acknowledge state array In RLC acknowledged mode, the BSN values specified in the RBB are interpreted by subtracting the bit position in the bitmap from the SSN modulo 128. A valid BSN value in the RBB is one that is in the range $[\ V(A) < BSN < V(S)\ ]$ *modulo* 128. These inequalities are interpreted in the following way: the BSN is valid if, and only if the following inequality holds:

$$[\ BSN - V(A)\ ]\ modulo\ 128 < [\ V(S) - V(A)\ ]\ modulo\ 128$$

Figure 3.18 illustrates the update process. At the RLC receiver, the starting sequence number (SSN) is assigned the value of the receive state variable $V(R)$. The RBB is assigned the 64 elements whose indices in the receive state array $V(N)$ are at the receiver range from $[\ V(R) - 1\ ]$ *modulo* 128 to $[\ V(R) - 64\ ]$ *modulo* 128. For each bit in the bitmap, the bit is assigned the value 1 if the corresponding element in $V(N)$ indexed relative to SSN has the value `RECEIVED`. The bit is assigned the value 0 if the element in $V(N)$ has the value `INVALID`.

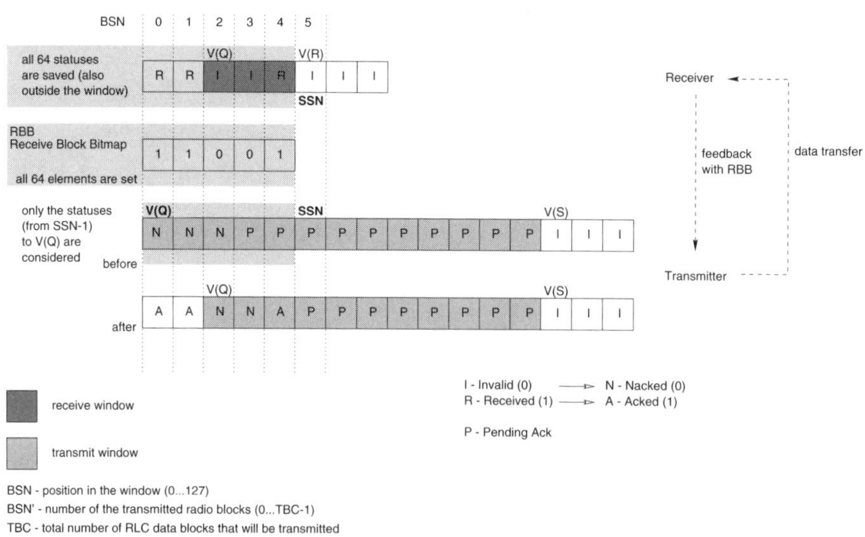

Figure 3.18: Status update of the transmit window using the RBB

At the RLC transmitter, each bit in the RBB whose corresponding BSN value is within the transmit window is considered. If the bit contains the value 1, the corresponding element in $V(B)$ indexed relative to SSN is set to the value `ACKED`. If the bit contains the value 0, and the instance of a timer (T3198) corresponding to BSN is expired, the element in $V(B)$ is set to the value `NACKED`. If the bit contains the value 0 and the instance of the timer is not expired (i.e., the RLC data block was recently (re)transmitted and thus can not be negatively acknowledged in this particular Packet Ack/Nack message), the element in $V(B)$ is not modified. A bit within the RBB whose corresponding BSN is not within the transmit window is ignored.

3.5.3.4 Segmentation Modes

The segmentation of LLC frames in GPRS Release 97/98 can be performed in two different modes: padding mode or concatenation mode (see Figure 3.19). From GPRS Release 99 on, only the concatenation mode is recommended.

Padding mode Using padding mode, only one LLC frame can be transported in one RLC data block, and if the length of the LLC frame does not fit into an integer number of RLC data blocks, the last RLC data block has to be filled up with padding bits. Therefore, transmission capacity is wasted. The implementation of this mode is rather simple in contrast to the concatenation mode.

Concatenation mode The concatenation mode provides the ability to transport several LLC frames within one single RLC data block. If the length of an LLC frame does not fit into an integer number of RLC data blocks, the free space in the last RLC data block can be used for the transmission of the next LLC frames.

If a consecutive LLC frame for instance is smaller than the remaining space in the actual RLC data block, it is possible to transmit the whole LLC frame in the RLC data block. The calculated free space can be used for the transmission of the beginning of the following LLC frame, and so on. Assuming a stream of LLC frames, the concatenation mode segments the transmission data into RLC data blocks without paying attention to the beginning or ending of the LLC frames.

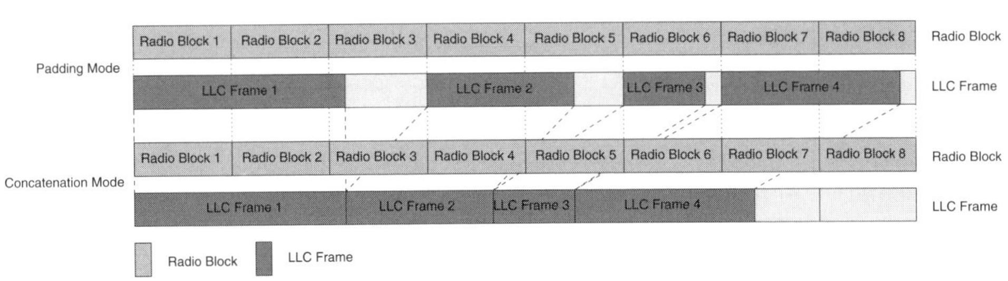

Figure 3.19: Segmentation of LLC frames

3.5.4 MAC Functions

The GPRS MAC layer is responsible for providing efficient multiplexing of data and control signaling on the uplink and the downlink. The multiplexing on the downlink is controlled by the downlink scheduler, which has knowledge of the active MSs in the system and of the downlink traffic. Therefore, an efficient multiplexing on the PDCHs can easily be made. On the uplink, the multiplexing is controlled by channel reservation to individual MSs. This is done by resource requests, which are sent by the MS to the network, which then has to schedule the uplink PDCHs. Additionally, a contention resolution is needed when several MSs attempt simultaneously to access the control channel for resource requests. Collision detection and recovery are included when using the Slotted Aloha reservation protocol for recovery, which means that the MS has to wait a random backoff time before a new access attempt is done.

The MAC procedures include the provision of a TBF, which allows a point-to-point transfer of data in one cell between the BSS and the MS. As a physical connection between two radio resource entities, the TBF is used to transport the RLC data blocks on the PDCHs. The TBF is maintained only for the duration of the data transfer and is released as soon as all data is sent (RLC send queue is cleared). A TFI is assigned to the TBF by the network to associate the MS with the current TBF.

An *Uplink State Flag* (USF) is used by the network to control the multiplexing of different mobile stations on uplink PDCHs. It is included in the header of each RLC/MAC PDU on a downlink PDCH. The USF indicates which station is allowed to transmit during the next uplink radio block period. The MS, which has the value indicated in the USF field, is allowed to transmit an RLC block in the uplink direction on the same PDCH, on which it has received the radio block with the corresponding USF. Note that the USF field has only 3 bit which means that only up to seven MSs can operate on one PDCH. One USF value is used to indicate the PRACH on the next uplink block period. Assuming a higher multislot capability, the MS occupies USFs on several PDCHs, which limits the total number of active MSs on one frequency. The USF points either to the next uplink radio block or to the sequence of four uplink radio blocks starting with the next uplink radio block (`USF Granularity`). The MS sends the `Packet Channel Request` in one of the four bursts on the radio block used for PRACH (indicated by setting USF=FREE). In the access the burst is selected randomly.

3.5.5 TBF Setup

Both the network and the MS can initiate the establishment of a *Temporary Block Flow* (TBF) on the PCCCH allocated in the cell. The access is carried out on the PCCCH in either one or two phases, whereas two-phase access is used if the requested RLC mode is unacknowledged mode to ensure a safe establishment or if more than one time slot is requested by the MS. A one-phase access procedure is used if the amount of data to send fits into eight or fewer RLC/MAC blocks using CS-1.

3.5.5.1 Uplink TBF Setup

The MS enters the packet access procedure by sending a `Packet Channel Request` message on the PRACH and entering the `packet transfer mode` (see Figure 3.20). This `Packet Channel Request` message contains parameters required to indicate the

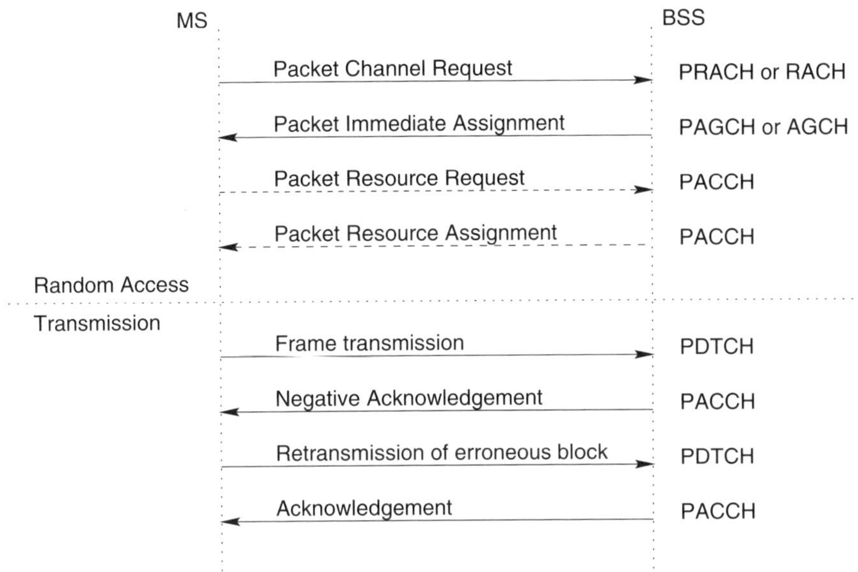

Figure 3.20: Uplink TBF establishment and data transmission

mobile station's request for radio resources and the type of access needed. Access persistence control on PRACH can be steered either by the network (network steered method) or by the mobile station (mobile station steered method) to avoid collision failure.

In one-phase access, the network responds to the `Packet Channel Request` with the `Packet Immediate Assignment` (`Packet Uplink Assignment`), reserving the resources on PDCHs for uplink transfer of a number of radio blocks.

In two phase access, the network responds to the `Packet Channel Request` with `Packet Immediate Assignment` (`Packet Uplink Assignment`), which reserves the uplink resources for transmitting the `Packet Resource Request`. The `Packet Resource Request` message carries the complete description of the requested resources for the uplink transfer. Thereafter, the network responds with a `Packet Resource Assignment` message (`Packet Uplink Assignment`), reserving resources for the uplink transfer.

If there is no response to the `Packet Channel Request` within a predefined time period (`T3168`), the MS retries after a random back-off time. On receipt of a `Packet Channel Request`, the network sends a `Packet Uplink Assignment` message on the same PCCCH on which the network has received the `Packet Channel Request` message.

Packet data traffic is bursty in nature. Sometimes, the BSS will receive more channel requests than it can serve within a certain time limit. To avoid the repeat of `Packet Channel Requests` the sender is notified with a `Packet Queueing Notification` that its message is correctly received and will be handled later.

Efficient and flexible utilization of the available spectrum for packet data traffic with one or more PDCHs in a cell can be obtained using a multislot channel reservation scheme. Blocks from one MS can be sent on different PDCHs simultaneously, thus reducing the packet delay for transmission across the air interface. The bandwidth may be varied by allocating one to eight time slots in each TDMA frame depending on the number of available PDCHs, the multislot capabilities of the MS, and the current system load.

As mentioned earlier, the master–slave channel concept requires mechanisms for efficient utilization of the uplink PDCHs. Therefore, the USF is used on PDCHs. The 3-bit USF, sent at the beginning of each radio block on the downlink, points to the next uplink radio block. It enables the coding of eight different USF states, which are used to multiplex the uplink traffic. The channel reservation command includes the list of allocated PDCHs and the corresponding USF state per channel. To an MS, the USF marks the part of the channel it can use for transmission. An MS monitors the USF and, according to the USF value, identifies PDCHs assigned to it and starts transmission. This allows efficient multiplexing of blocks from a number of MSs onto a single PDCH. Additionally, the channel reservation command can be sent to the MS even before the total number of requested PDCHs is free. Thus, the status flags not only result in a highly dynamic reservation but also allow interruption of transmission because of a pending message or high-priority messages. One USF value is used to denote PRACH (`USF=free`). The other USF values (`USF=R1/R2/···/R7`) are used to reserve the uplink for different MSs.

3.5.5.2 Downlink TBF Setup

A BSS initiates a packet transfer by sending a `Packet Paging Request` on the PPCH on the downlink in order to determine the location of an MS (if it is not already known). The MS responds to the paging message by initiating a procedure for page response very similar to the packet access procedure described earlier. If the location is known, the paging procedure is followed by the `Packet Resource Assignment` for downlink frame transfer containing the list of PDCHs to be used. The last procedure is also performed before downlink transfer, if the location of the MS is already known.

Since an identifier (e.g., TFI) is included in each radio block, it is possible to multiplex radio blocks destined for different MSs on the same downlink PDCH. It is also possible to interrupt a data transmission to one MS if a higher-priority data or pending control message is to be sent to some other MS. Furthermore, if more than one PDCH is available for downlink traffic, and provided the MS is capable of monitoring multiple PDCHs, blocks belonging to the same flow can be transferred on different PDCHs in parallel.

3.5.6 RLC Block Transfer

3.5.6.1 Uplink

When the MS receives the complete uplink assignment, it begins to monitor the assigned PDCHs for the USF value. If there is already a TBF running, the MS waits for the moment of the TBF starting time, which is specified in the `Packet Uplink Assignment` message. Then the MS starts to use the new parameters. Otherwise, if there is no TBF running, the MS begins to monitor the PDCH for USFs as soon as the starting time expires.

It is possible to set the `RLC-DATA-BLOCKS-GRANTED` information element in the `Packet Uplink Assignment` message to allow the MS to send only a specified number within the TBF.

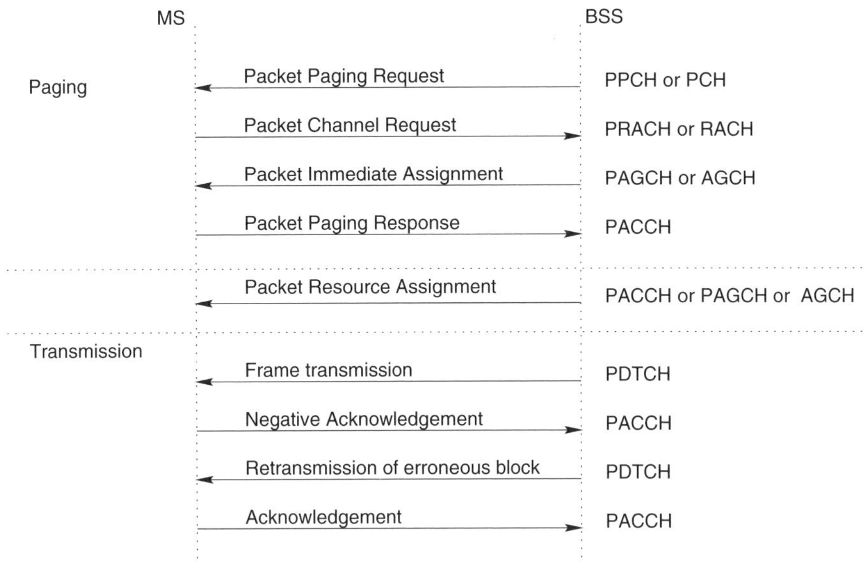

Figure 3.21: Downlink TBF establishment and data transmission

When transmitting RLC blocks on the uplink, the first three RLC data blocks of the uplink TBF contain a TLLI field in the RLC data block header. Each time the MS detects an assigned USF value on an assigned PDCH, one RLC block will be transmitted on the same PDCH in the next block period. When an RLC block is transmitted to the network, the MS starts the timer `T3180`. When the assigned USF value is detected, the MS resets `T3180`. On expiry of timer `T3180`, the MS performs the abnormal release with random-access procedure.

On the network side, the counter `N3101` will be reset if a valid RLC block arrives, but it will be incremented for each radio block for which no data is received. If the counter reaches its maximum value, the network starts timer `T3169`. After its expiry, the network may reuse the USF and TFI for another TBF.

3.5.6.2 Downlink

After the reception of the `Packet Downlink Assignment` message, the MS starts the timer `T3190` to define when to stop waiting for the valid data from the network side following the initial `Packet Downlink Assignment`.

If there is no downlink TBF in progress and the received message contains a TBF starting time information element, the MS remains on the PCCCH until the TDMA frame number, indicated by the TBF starting time, arrives. Now the MS begins to monitor the assigned downlink PDCHs.

If there is already a downlink TBF in progress, the MS continues as before until the indicated TDMA frame number occurs, at which the MS begins to use the new assigned downlink TBF parameters.

Receiving a valid RLC data block without the FBI bit set to 1, the MS resets and restarts timer `T3190` to define when to stop waiting for valid data from the network. Otherwise (`FBI=1`) the timer `T3190` will be stopped. In this case the TBF release

Table 3.2: Coding parameters

	CS-1	CS-2	CS-3	CS-4
Code rate	1/2	2/3	3/4	1
GPRS net data rate [kbit/s]	9.05	13.4	15.6	21.4
RLC data block size [bit]	176	263	307	423
MAC header size (ex. USF) [bit]	5	5	5	5
RLC/MAC block size (ex. USF) [bit]	181	268	312	428
USF [bit]	3	3	3	3
BCS [bit]	40	16	16	16
Precoded USF [bit]	3	6	6	12
Tail bits	4	4	4	0
Radio block [bit] (before conv. coding)	228	294	338	456
Convolutional code rate	1/2	1/2	1/2	1
Coded radio block size [bit]	456	588	676	456
Puncturing [bit]	0	132	220	0
Radio block size [bit]	456	456	456	456

procedure is triggered and no retransmissions are made. On expiry of T3190 (TBF release procedure failed), the TBF will be terminated with abnormal release.

3.5.7 Channel Coding Schemes

The redundancy added to the RLC/MAC blocks can be adapted by the operator to the channel conditions in the network. The GPRS channel *Coding Schemes* (CSs) CS-1 to CS-4 enable code rates from 1/2 to 1 (see Table 3.2). The RLC data block size depends on the number of LLC PDUs concatenated in the radio block. In this table one LLC PDUs per radio block is assumed. CS-1 has to be used for RLC/MAC control blocks after the GPRS specification. Although all CSs can be used for RLC/MAC data blocks, only CS-2 is used for RLC/MAC data blocks in initial GPRS deployments.

In the first step of the coding procedure a BCS is added for error detection. The second step consists of precoding the USF (except for CS-1), adding four tail bits, convolutional coding for error correction and puncturing to realize the desired code rate. The result is a bit pattern with a length of 456 bit independent of the coding scheme used. For CS-4 no tail bit is added and no convolutional coding (and puncturing) is performed; only the USF is precoded. For information about puncturing see [17].

In GPRS the punctured convolutional code of the GSM 1/2-convolutional coding is used (see Section 2.1.7.5).

The decoding of the USF is simplified whereby for coding CS-2–4 a 12-bit long code word, which is not punctured, is generated for the USF. With CS-2 and CS-3 the USF is precoded with 6 bit before the frame is convolutionally coded, with the first 12 bit not being punctured. When CS-1 is used, the entire frame is coded and the USF has to be decoded as part of the information field. Figure 3.22 shows an example of the coding of an information frame for CS-3.

The GPRS *net data rate* R_{net} is defined by the data rate of RLC/MAC blocks considering the channel coding overhead and considering the idle frames. Since one 52 multiframe comprises 12 radio block periods, the mean transmission duration of a radio block D_{RB} can be calculated as:

$$D_{\text{RB}} = \frac{52\ frames \cdot 4.615\ ms}{12\ frames} = 20\ ms \tag{3.1}$$

With X as the number of bits in one RLC/MAC block from Table 3.2 (RLC/MAC block size) the net data rate can be calculated as:

$$R_{\text{net}} = \frac{X\ bit}{D_{\text{RB}}} \tag{3.2}$$

For example:

$$R_{\text{net(CS-1)}} = \frac{181\ bit}{20\ ms} = 9.05\ \frac{kbit}{s} \tag{3.3}$$

Note that the net data rate in GPRS does not equal the data rate that the RLC/MAC layer offers to the LLC layer. The latter can be calculated based on the number of RLC data block information bytes from Table 3.1.

3.6 Physical Layer (PL)

The *Physical Layer* (PL) at the air interface is divided into the *Physical Link Layer* (PLL) and the *Radio Frequency* (RF) layer. These sublayers are specified in [13, 14]. While in the RF sublayer mainly modulation and demodulation are carried out (see Section 2.1.7.5), the PLL sublayer provides services for data transmission over the radio interface. The PLL sublayer is responsible for the FEC coding, allowing the

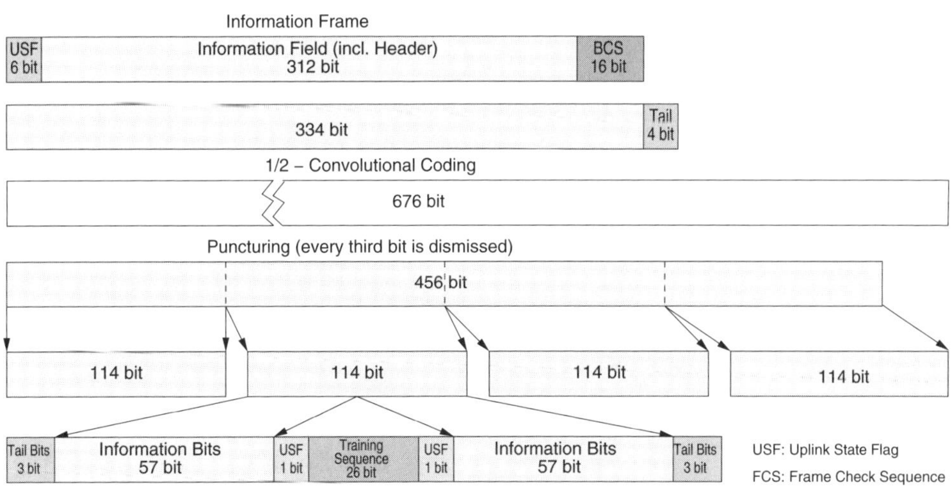

Figure 3.22: Coding of an information frame according to CS-3

detection and correction of erroneous transmitted code words and the indication of uncorrectable code words. It is also responsible for the interleaving of one radio block over four bursts in consecutive TDMA frames and provides procedures for synchronization.

4
GPRS Signaling

The GPRS protocol architecture is organized into two planes, the user plane, also called transmission plane, and the control plane, also called signaling plane. The user plane is responsible for data transmission, when packet data transfer is actually requested in uplink or downlink. To realize the transfer between the correct network nodes with adequate performance characteristics, the user plane protocols need certain information such as addresses of peer entities, the status of network elements or requested protocol options. This information has to be provided at the beginning of a session and has to be kept up-to-date during an ongoing session, while the mobile user might move around with his terminal or use different services in parallel or consecutively with idle periods in between.

The GPRS control plane realizes these management functions, namely *GPRS Mobility Management* (GMM), *Session Management* (SM) and *Quality of Service* (QoS) management. GMM keeps track of the MS's location and its state, updates databases with this information and supports cell change procedures. The SM manages the logical context between the MS and an external *Packet Data Network* (PDN) based on TCP/IP or X.25, when a new session is set up or changing its performance requirements. In GPRS networks different applications with different performance requirements are supported. QoS management in GPRS networks enables the network to establish QoS contracts with the MS differentiating applications and subscribers according to their QoS requirements. The GPRS QoS support is based on the SM procedures.

4.1 The GPRS Control Plane

In Figure 4.1 the GPRS control plane between MS and SGSN is shown. The signaling layers SM and GMM reuse the layer-2 protocols and the GPRS channel interface of the user plane (see Chapter 3) to transmit signaling messages over the air interface. The control plane between SGSN and the registers shown in Figure 4.2 (see Section 2.1.3) is based on the GSM-specific extension of the Signaling System Number 7 called *Mobile Application Part* (MAP) and an enhanced version of the *Base Station Subsystem Application Part* (BSSAP), called BSSAP+ at the interfaces G_r (SGSN-HLR), G_s (SGSN-MSC/VLR) and G_f (SGSN-EIR). In the GPRS core network between the *GPRS Support Nodes* (GSNs) the control part of the GTP, the *GPRS Tunneling Protocol (Control plane)* (GTP-C) (see Section 3.1) is used to perform signaling in the core network (see Figure 4.3).

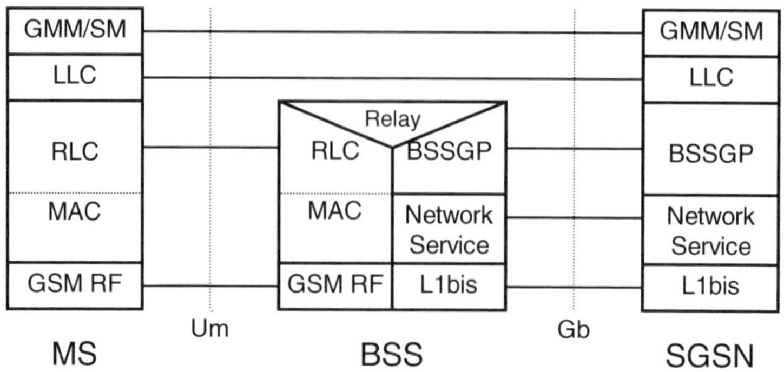

Figure 4.1: GPRS control plane (MS - SGSN)

4.2 Mobility Management

Mobility management in GSM and GPRS [29] includes functions for location registration with the PLMN and location updating to report the current location of an MS. Additionally it is responsible for identification and authentication of subscribers.

The GMM concept allows combined procedures for GSM and GPRS to reduce the signaling load. The handover procedure is split up into the cell update and the *Routing Area* (RA) update.

In contrast to the GSM system, where the handover decision is made by the network based on measurement reports sent by the MS [2], the cell change, also called cell reselection, in GPRS is not necessarily prepared and triggered by the network, but is mostly decided by the MS, when it detects that it has entered a new cell or even a new RA by reading the system information on the broadcast channel.

4.2.1 Location Area and Routing Area

The *Location Area* (LA) and *Routing Area* (RA) are different areas comprising a certain number of cells (see Figure 4.4) [86]. For example one LA may comprise 30 cells. This hierarchical structure with LAs is chosen for GSM networks to manage the location of an MS. The network initially only knows the LA in which the MS is located. The MS provides this information periodically, or when the MS changes its LA. When it becomes necessary to determine the exact cell location of an MS, the network sends a paging command in all cells of the LA. The MS, which continuously monitors the broadcast channel, responds to this paging command so that a connection or a context can be established. With this procedure the signaling load on the air interface is minimized, since the network need not know about the exact cell location of the MS when it is not active. Additionally, the VLR does not have to be updated by the SGSN, as long as the MS is staying inside one LA.

The RA that does not exist in GSM networks without GPRS support was introduced to speed up the paging procedure and lower the signaling load for paging. The number of cells in an RA is generally less than or equal to that of an LA. For example, one LA with 30 cells may comprise three RAs with 10 cells each. An RA can not span more than one LA and an RA is served by only one SGSN. One SGSN can handle several RAs and the size of an RA can range from a part of a city to an entire

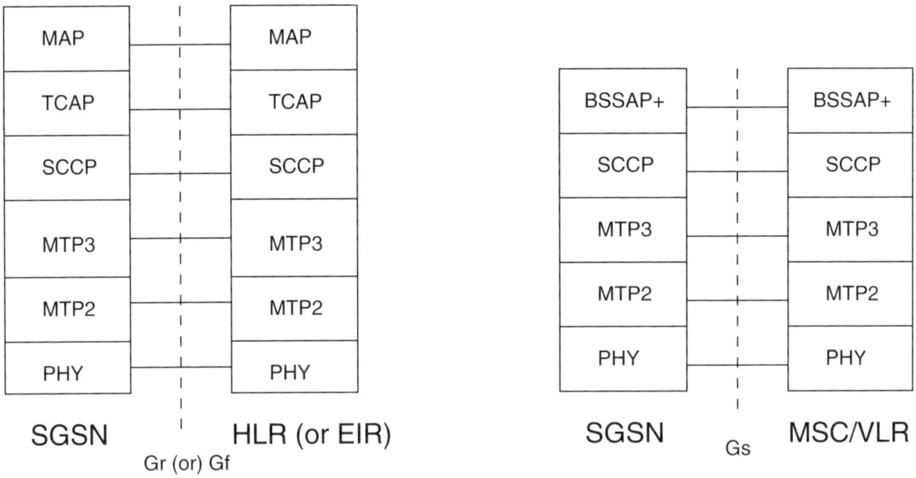

MAP Mobile Application Part
TCAP Transaction Capabilities Application Part
BSSAP+ BSS Application Part +
SCCP Signaling Connection Control Part
MTP Message Transfer Part

Figure 4.2: GPRS control plane (SGSN - HLR, SGSN - VLR)

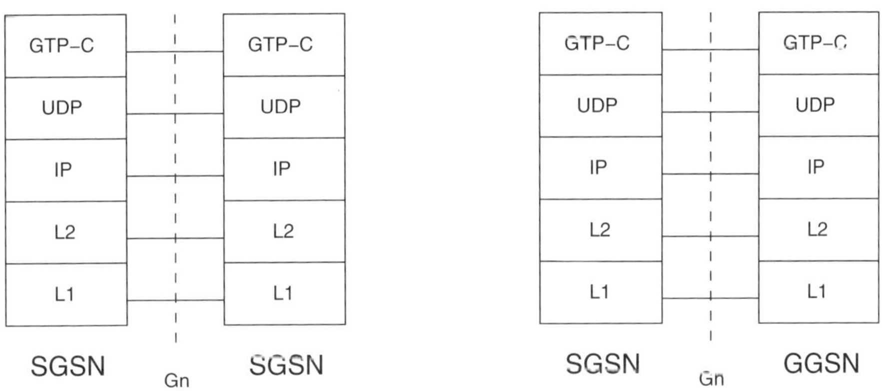

Figure 4.3: GPRS control plane (SGSN - SGSN, SGSN - GGSN)

province or even a small country. Figure 4.4 shows an example cell scenario, in which it is possible to see how cells could be grouped in a cell planning configuration. Cells 1, 2 and 3 together build one RA. Cell 4 alone builds another distinct RA. Together these two RAs form one LA. A set of LAs are served by one BSC, and many BSCs are served by an SGSN. The SGSNs are interconnected and represent the highest level in the GMM hierarchy. Cell 14 belongs to another SGSN in this example.

After an MS has changed from one cell to another it reads the system information of the new cell. Since the RA is broadcast as part of the system information, the MS can determine whether it has changed its RA. If this is the case, the MS sends an RA update to the SGSN to inform it about the new RA. The cell update and RA

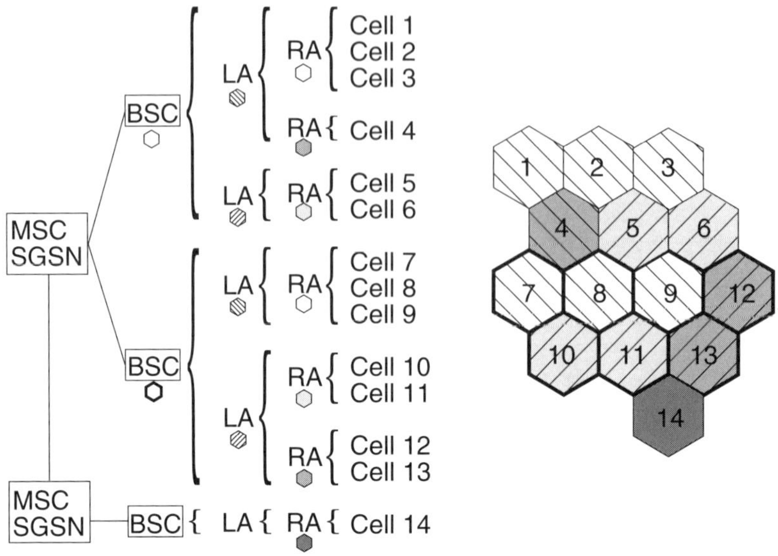

Figure 4.4: Example of a cell scenario comprising all cell change possibilities

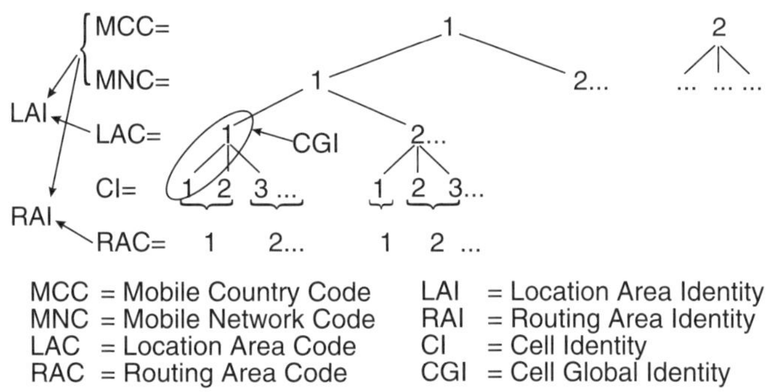

Figure 4.5: Relationship of the different GMM identities in GPRS

update procedures are described in detail in Section 4.2.3. The following rules for the different identities apply in GPRS (see Figure 4.5):

- The *Location Area Code* (LAC) identifies an LA within one GSM network. The *Location Area Identity* (LAI) is unique among all networks and is composed of the *Mobile Country Code* (MCC), the *Mobile Network Code* (MNC) and the LAC. The MCC identifies the country in which the GSM PLMN is located.
- The *Routing Area Code* (RAC) identifies the RA inside one LA and is only unique when presented together with the LAI. Therefore the *Routing Area Identity* (RAI) is composed of the LAC and the RAC.

- The *Cell Identity* (CI) identifies a radio cell inside one LA and is only unique when presented together with the LAI. A unique cell identifier, the *Cell Global Identity* (CGI), is composed of LAI and CI [4].

4.2.2 Mobility Management Procedures

The *GPRS Mobility Management* (GMM) procedures comprise access control and authentication functions. To obtain access to GPRS services the MS has to initiate the GPRS attach procedure (see Figure 4.6). During *Mobility Management* (MM) procedures user data can, in general, be transmitted while the signaling is going on. This can lead to loss of data during attach, authentication and RA update procedures. In order to minimize retransmissions the MS and SGSN should not transmit user data during these procedures.

4.2.2.1 GPRS Attach Procedure

During the GPRS attach procedure the MS communicates with the SGSN. The MS provides its *International Mobile Subscriber Identity* (IMSI) (see Section 2.1.5) or *Packet Temporary Mobile Subscriber Identity* (P-TMSI), if available, for identification at the SGSN in the `Attach Request` message. Different types of attach procedures are defined in [29] depending on which identifiers are available before the attach request. At the RLC/MAC layer, the MS shall identify itself with a Foreign TLLI, or a Random TLLI (see Section 3.4.2) if a valid P-TMSI is not available. The Foreign or Random TLLI is used as an identifier during the attach procedure until a new P-TMSI is allocated.

The parameters of the `Attach Request` message include the MS's GPRS multislot capabilities, supported GPRS ciphering algorithms, the type of attach performed, and further modes and capabilities of the MS.

If the MS identifies itself with P-TMSI and the SGSN has changed since detach, the GTP-C entity (Section 4.1) of the new SGSN sends an `Identification Request` message to the old SGSN to request the IMSI. The old SGSN responds with `Identification Response` containing the IMSI and authentication parameters. If the MS is unknown in both the old and new SGSN, the GMM entity of the SGSN sends an `Identity Request` to the MS. The MS responds with `Identity Response` and provides the IMSI. After authentication and security functions are performed between SGSN and the GPRS registers using information requested from the MS, the location of the MS is updated with the messages `Update Location` and `Update Location Achnowledge` between SGSN and HLR or VLR using MAP and BSSAP+ (see Section 4.1) signaling, respectively. After that the GMM entity of the SGSN sends an `Attach Accept` message with the assigned P-TMSI and the VLR TMSI to the MS. If the P-TMSI or VLR TMSI was changed, the MS acknowledges the received TMSI returning an `Attach Complete` message to the SGSN. If the VLR TMSI was changed, the BSSAP+ entity of the SGSN confirms the VLR TMSI reallocation by sending a `TMSI Reallocation Complete` message to the VLR. If the `Attach Request` cannot be accepted, the GMM entity of the SGSN returns an `Attach Reject` message to the MS.

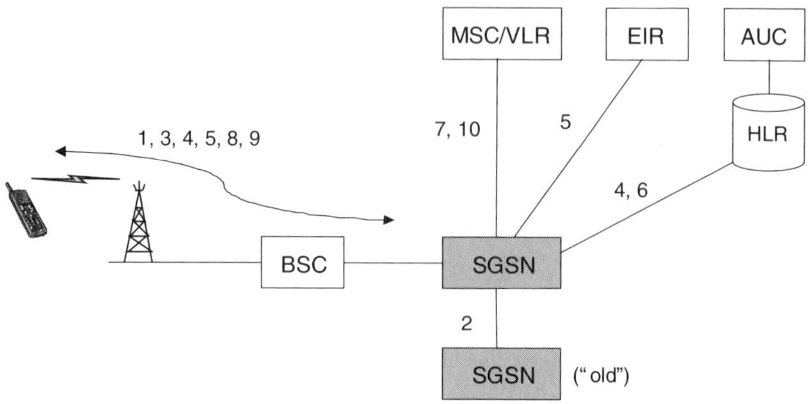

1. Attach Request (MS - SGSN)
2. Identification Request (new - old SGSN)
3. Identity Request (SGSN - MS)
4. Authentication (SGSN - HLR, SGSN - MS)
5. IMEI Check (SGSN - MS, SGSN - EIR)
6. Update Location (SGSN - HLR)
7. Location Updating (SGSN - MSC/VLR)
8. Attach Accept (SGSN - MS)
9. Attach Complete (MS - SGSN)
10. TMSI Reallocation Complete (SGSN - MSC/VLR)

Figure 4.6: Overview of the attach procedure in GPRS

4.2.2.2 GPRS Detach Procedure and Purge Function

The GPRS detach procedure allows an MS to inform the network that it does not want to access the GPRS services any longer and allows the network to inform an MS that it does not have access to the SGSN-based services any more. The MS is detached either explicitly or implicitly. In the explicit detach procedure the network or the MS explicitly requests detach, while in the implicit detach, the network detaches the MS, without notifying the MS, a configuration-dependent time after the mobile reachable timer expired, or after an irrecoverable radio error causes disconnection of the logical link. In the explicit detach case, a `Detach Request` is sent by the SGSN to the MS, or by the MS to the SGSN. If the MS performs the detach procedure, it sends a `Detach Request` message to the SGSN. If requested by the MS a `Detach Accept` message is returned.

The *purge function* allows an SGSN to inform the HLR that it has deleted the *Mobility Management* (MM) and *Packet Data Protocol* (PDP) contexts (see Section 4.3) of a detached MS. The SGSN may, as an implementation option, delete the contexts of an MS immediately after the implicit or explicit detach of the MS. Alternatively, the SGSN may keep for some time the contexts of the detached MS, so that they can be reused at a later GPRS attach without accessing the HLR. After deleting the contexts of a detached MS, the SGSN sends a `Purge MS` message to the HLR containing the IMSI. The HLR updates the entry of the MS in the database and acknowledges with a `Purge MS Acknowledge` message.

4.2.2.3 Authentication

The security authentication function for prevention of unauthorized GPRS service usage and confidentiality is performed during the attach procedure (see Sec-

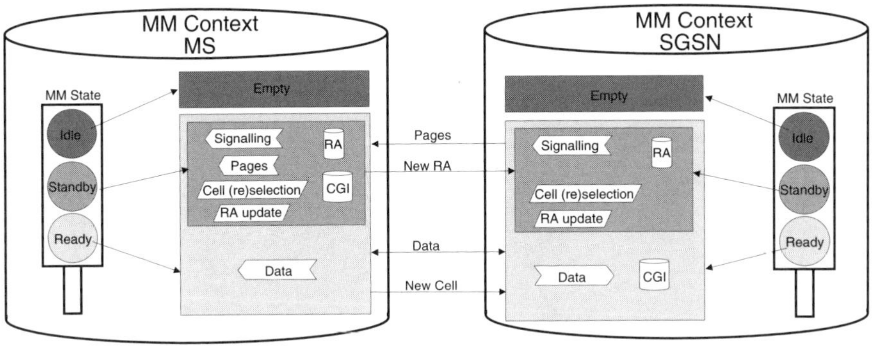

Figure 4.7: Mobility management states on the MS and SGSN side

tion 4.2.2.1). This function performs the identification and authentication of the service requester, and the validation of the service request type to ensure that the user is authorised to use the particular network services. Security-related network functions are completely based on the GSM standard as described in [1].

4.2.2.4 States of the Mobility Management

The GMM functionalities are based on three different MM states defined at the MS and the SGSN states, in which each MS can be. These MM states are:

1. The `idle` state.
2. The `standby` state.
3. The `ready` state.

In each of these MM states the MS and the SGSN hold a different set of information about the GPRS terminal. These are denoted MM contexts. Figure 4.7 shows the different states on the two sides. The MM context in the MS comprises among others the MM state MS identifiers, location and security data and zero or more PDP contexts, while the MM context at the SGSN additionally contains classmarks for the communication with the registers.

In the idle state the MM context is empty, because the subscriber is not attached to the GPRS network. The MS and SGSN do not hold any valid information about the cell and RA in which the MS is located. Therefore no MM procedures are performed. The MS is seen as not reachable and can not be paged.

In the standby state the MS is attached to the network. Each of the MS and SGSN have established an MM context. Both the MS and the SGSN know the RA in which the MS is located. In this state the SGSN can send paging messages to all the cells of the RA to find the MS, if it wants to send messages to it in the downlink direction. The MS may receive PTM-M and *PTM Group Call* (PTM-G) paging messages, but can not receive or transmit any data.

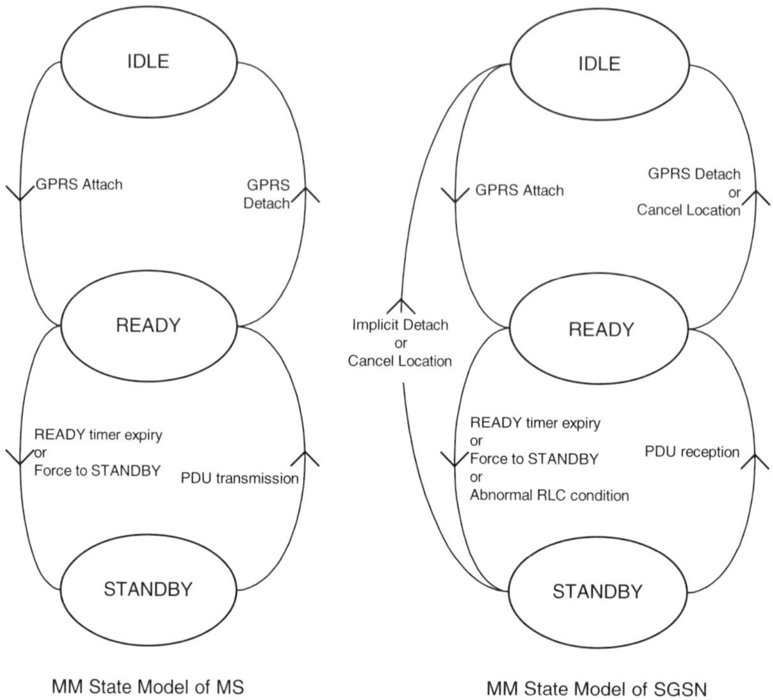

Figure 4.8: MM state transitions

In the ready state the SGSN knows in which cell the MS is located and can therefore send a continuous stream of data packets on the downlink, even if the MS changes its cell supported by *Location Management* (LM) procedures (see Section 4.2.3). The MM context is practically the same as in the case of the `standby` state but extended by the information on the cell in which the MS is located. The MS may receive PTM-M and PTM-G data (see Section 2.2.3) in the `ready` state and the SGSN must not initially page the MS to send data. In this state the MS can activate or deactivate PDP contexts. The MM context remains in the `ready` state even if there is no data to send. It returns to the `standby` state after the expiry of the `ready` timer (see Section 4.2.2.6).

4.2.2.5 GMM State Transitions

In Figure 4.8 all possible state transitions of the MM context are shown. For example, when moving from `idle` to `ready` the MS performs a GPRS attach and a logical link is initiated to an SGSN. When the implicit detach occurs the MM and PDP contexts in the SGSN return to `idle` and `inactive` state and the GGSN PDP context is deleted. A cancel location is triggered when a MAP cancel location message is received from the HLR. When an MS sends a PDU to the SGSN its MM state switches to `ready` and when the SGSN receives this PDU it also goes into the `ready` state. The SGSN can also force the state to `standby` before the `ready` timer expires. The context returns also to `standby` after an irrecoverable radio error causes disconnection of the logical link.

4.2.2.6 GMM Timers

Besides the events described above that are controlling the state transitions, timers are defined in the GMM that can also provoke state transitions.

The ready timer controls the duration an MM context remains in the `ready` state in the MS and in the SGSN. This timer is reset and begins running on the MS side when an LLC PDU is transmitted and on the SGSN side when an LLC PDU is correctly received. At expiry of the timer the context on both sides returns to `standby`. The duration of the timer is the same on both sides and only the SGSN can change its value using `Attach Accept`, `RA Update Accept` or `PDP Context Accept` messages. If the `ready` timer length is set to zero, the MS immediately is forced into `standby` state. If the timer length is set to 1 s, the `ready` timer function is deactivated, i.e. the timer no longer runs and the MS remains in `ready` state.

The Periodic RA Update timer is set periodically and its length, which is constant in one RA, is included in the `RA Update Accept` message from the SGSN or after an `Attach Accept` message. At expiry the MS performs an RA update procedure.

The Mobile reachable timer controls the periodic RA update procedure in the SGSN. It is slightly longer then the periodic RA update timer. It is set when the SGSN context returns from `ready` to `standby` and stopped when the transition to `ready` is performed. This timer is reset when a periodic RA update reaches the SGSN, because in this case, the SGSN knows that the MS is reachable. If the timer expires then the MS is, e.g., out of coverage and the SGSN stops paging the MS.

4.2.3 Location Management Procedures

The *Location Management* (LM) function controls the cell and PLMN selection and provides a mechanism which enables the SGSN to know the RA of the MS in the `standby` and `ready` states [29]. If an MS is in the `ready` state, the SGSN knows the location of the MS on cell level based on the CGI.

Figure 4.9 shows a node view of the GMM, which is located in the SGSN and communicates over the LLC layer with the MS that this SGSN serves. The LLC PDUs are transmitted over the RLC/MAC layer across the air interface to the MS. The GMM has to maintain that the packet flow is routed to the correct destination, i.e. first to the correct BSC, then to the correct BTS.

When an MS enters a new cell, three different procedures can be performed:

1. A cell update procedure.

2. An RA update procedure (intra- or inter-SGSN).

In Figure 4.10 all the cell change possibilities are shown. These can be derived from Figure 4.4, in which a cell planning scenario is shown. All these cases will be described below in detail. The first possibility is the intra-SGSN cell update intra-BSC where an MS changes between two cells belonging to the same RA served by one BSC. In this case, the MS performs a cell update procedure. The second possibility is the

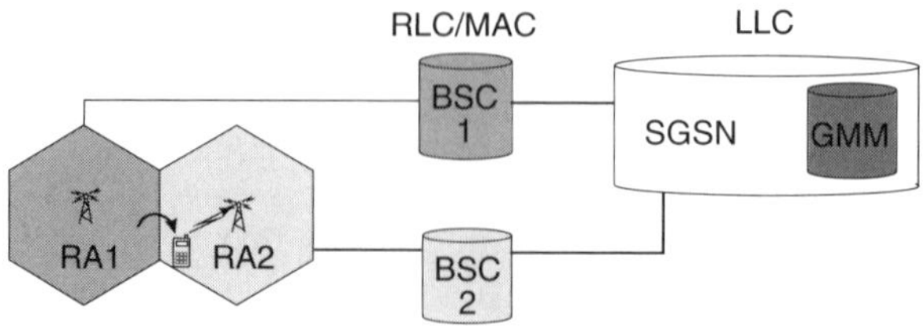

Figure 4.9: Node view of the GPRS mobility management on the network side

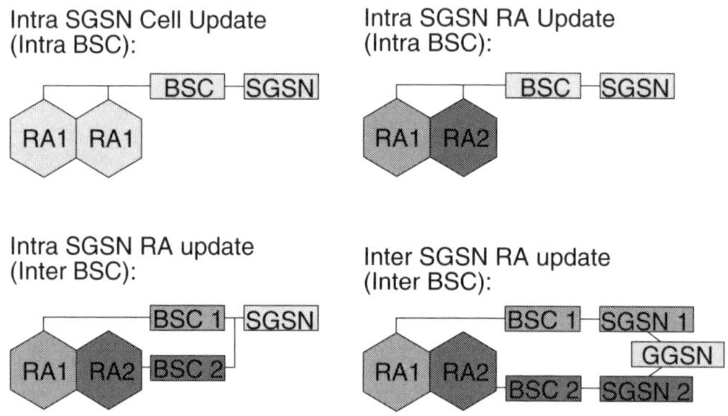

Figure 4.10: All cell change possibilities in GPRS

intra-SGSN RA update intra-BSC. Here the RA changes from one cell to the other. The MS has to perform an RA update procedure. In the third case the responsible BSC changes (intra-SGSN RA update inter-BSC). The fourth and last case is the inter-SGSN RA update where also the responsible SGSN changes.

The cell update and the RA update are mostly triggered by the MS in GPRS networks. This differs from the procedure in circuit-switched GSM networks, where the handover is completely controlled by the network. The reason for this difference is that GPRS is designed for packet transport, where delays due to a cell change are not as critical as delays during a voice conversation. In the GSM the handover is prepared by the network in a way which allows a conversation to stay uninterrupted and undisturbed. In GPRS there is a clear interruption of the service due to the fact that the network has first to be informed about the cell change and in the case of an RA update the network has to permit the MS to use the new cell it just entered.

In the following sections the cell update and the RA update are described.

4.2.3.1 Cell Change Signaling

Figure 4.11 shows a simplified overview of the cell change procedures in GPRS. Basically, two cell change possibilities are foreseen. The first one (1-a) is the cell update (see Section 4.2.3.2), where the MS notifies the SGSN GMM about a cell change by sending an uplink PDU in the new cell it enters. The second one (1-b) is the RA update (see Section 4.2.3.3), where the MS sends an `RA Update Request` message in the new cell it enters and must wait for a response from the GMM. The response

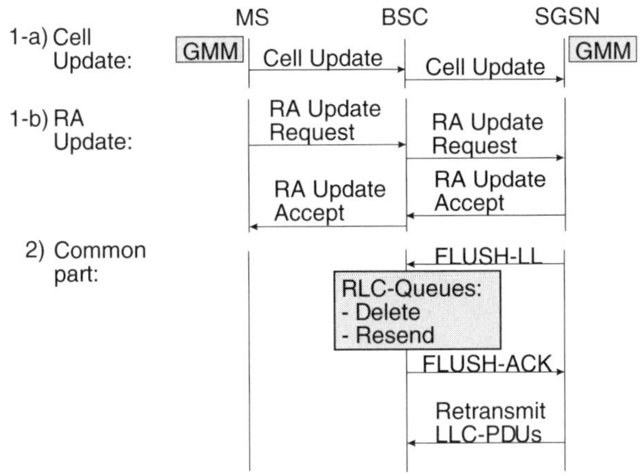

Figure 4.11: Message sequence chart of the cell change procedure

shown in Figure 4.11 is an `RA Update Accept` message, but it also might be an `RA Update Reject` message.

Then, there is a common part (2) for both the cell update and the RA update where the GMM decides with the `FLUSH-LL` message to delete or to forward RLC PDUs queued in the old BSC [15]. This message is acknowledged with the message `FLUSH-ACK`. After this response the SGSN can retransmit the LLC-PDUs immediately, if they have been deleted at the BSC.

In the following sections the different cases for cell update and RA update are explained in detail.

4.2.3.2 GPRS Cell Update

A cell update takes place when the MS enters a new cell inside the current RA and the MS is in `ready` state. If the RA has changed, a routing area update is executed instead of a cell update.

The MS performs the cell update procedure by sending an uplink LLC frame of any type containing the MS's identity to the SGSN. In the direction towards the SGSN, the BSS adds the CGI including RAC and LAC to all BSSGP frames [15]. A cell update is any correctly received and valid LLC PDU carried inside a BSSGP PDU containing a new CGI.

The SGSN detects this cell change, and the user data traffic for the MS is directed to the new cell.

Figure 4.12 shows what happens first, when a GPRS attached MS changes its cell while transmitting or receiving user data. Before the cell change an RLC/MAC TBF (see Section 3.5) in downlink is used to transmit RLC PDUs using the ARQ window mechanisms. Between LLC and RLC the BSSGP protocol is used. After the MS has changed the cell, it changes to `packet idle mode` (1) and monitors the PBCCH to read the system information of the new cell (3). In the meantime some RLC packets are still transmitted to the old cell (4), where they are lost (5). For an explanation of the RLC/MAC states `packet idle mode` and `packet transfer mode` see Section 3.5.1.

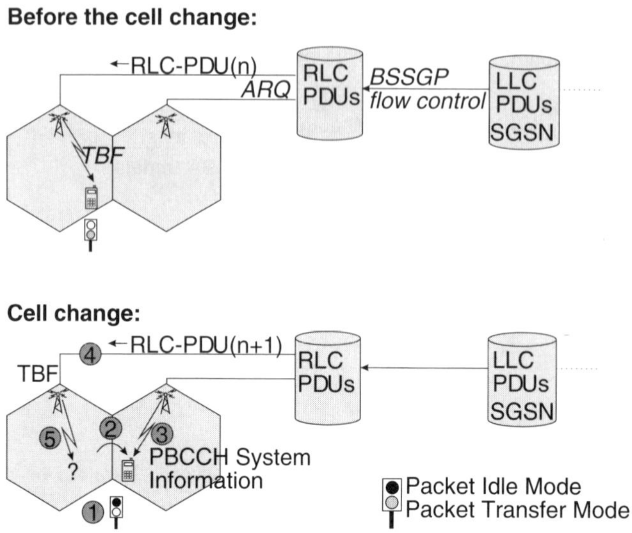

Figure 4.12: Common procedure for different cell change procedures

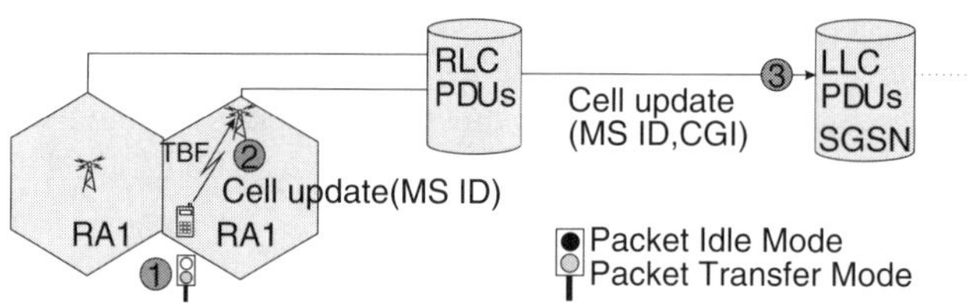

Figure 4.13: Cell update: uplink procedure

In Figure 4.13 the uplink phase of the cell update signaling is shown. The MS GMM sends an empty frame or a pending LLC frame containing the MS identity (2) after establishment of a new TBF (1). The GMM sends first the LLGMM-TRIGGER-REQ message to LLC after having restarted the GMM `ready` timer. Then the LLC protocol entity transmits a message to the GMM SGSN (see Section 3.4.3). If no LLC PDUs are pending and LLC is in the ABM mode, a supervisory frame is sent. If LLC PDUs are pending, an information frame or unnumbered information frame is sent depending on the actual state of the LLC entity that can be ABM or ADM, respectively (see Section 3.4). The BSC adds the CGI to the LLC frame (3), which is transmitted to the SGSN, where the GMM detects the cell change.

The SGSN does not send any response to the MS. As shown in Figure 4.14, the SGSN sends instead a FLUSH-LL message to the same BSC from which the LLC frame was received. If this message contains both the old cell identifier and the new one, then the BSC re-sends all its pending LLC frames to the new cell (Packet diversion in Figure 4.14). If the new cell identifier is not included within the FLUSH-LL message, the RLC PDUs in the BSC are deleted (Packet deletion in Figure 4.14). This case is foreseen in the standard, but only makes sense in practice if the BSC does not support the re-sending of LLC frames to a new cell.

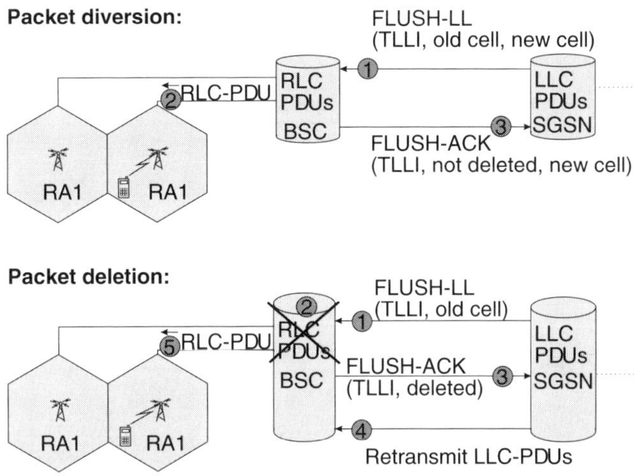

Figure 4.14: Cell update: downlink procedure

In both cases the BSC answers with the FLUSH-ACK message to notify whether the packets have been deleted or not. This enables the SGSN to retransmit the LLC PDUs if they have been deleted. The retransmission can either be performed directly or LLC can rely on the LLC retransmission mechanism, if LLC is operating in the acknowledged mode.

4.2.3.3 GPRS Routing Area Update (Intra-SGSN)

An MS can move between different RAs within the same PLMN by performing an RA update procedure. The RA update can be an Intra-SGSN RA update, where the new RA is served by the same SGSN or an Inter-SGSN RA update, where the new RA and the old RA are served by different SGSNs.

Intra-SGSN RA update To initiate this procedure the MS GMM entity sends an `RA Update Request` to the SGSN GMM. This message includes the identifier of the old RA and the old P-TMSI signature. The BSS adds the cell identity where the message was received before passing the message to the SGSN. Then security functions may be executed. If due to regional subscription restrictions the MS is not allowed to attach in the RA or subscription checking fails, the SGSN rejects the RA update. If the subscription and security checking is successful, the SGSN updates the MM context for the MS. A new P-TMSI may be allocated. An `RA Update Accept` is returned to the MS including the P-TMSI. If the P-TMSI was reallocated, the MS acknowledges the new P-TMSI with a `RA Update Complete` message (see Figure 4.15).

In the RA update procedure the new and the old RA can be served by the same or by two different BSCs. These two different cases are described in the following.

Routing Area update intra-BSC Figure 4.16 shows the uplink signaling procedure after the MS has moved to a new RA served by the same BSC. The MS sends the `RA Update Request` message (2) over the BSC to the SGSN after establishing a new

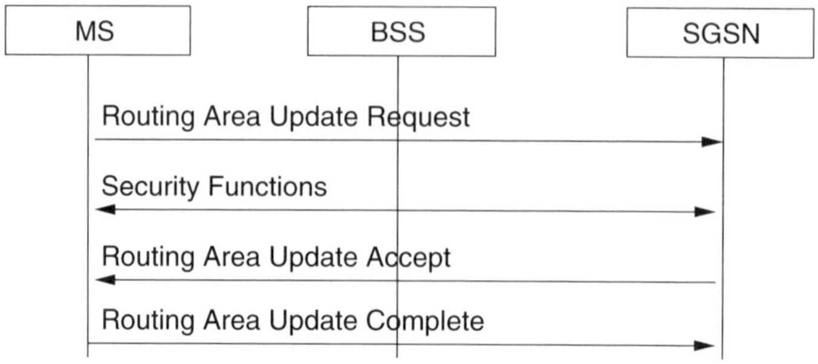

Figure 4.15: Intra-SGSN RA update

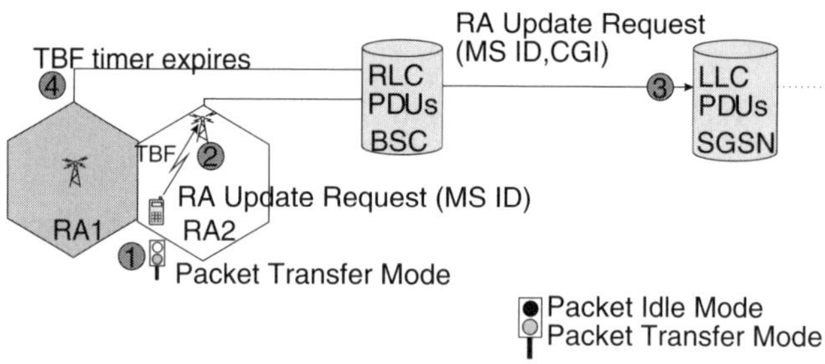

Figure 4.16: RA update Intra-BSC: uplink procedure

RLC/MAC TBF and going into `packet transfer mode` (1). The `RA Update Request` message is forwarded to the SGSN after the BSC has added the CGI (3).

The downlink part is shown in Figure 4.17. After the `RA Update Accept` message (1) is sent by the SGSN, the same LLC frame retransmission procedure as described in Section 4.2.3.2 is performed using `FLUSH-LL` messages (2, 3, 4, 5).

Routing Area update inter-BSC The special case of the RA update inter-BSC is shown in Figure 4.18. After the MS switches to the `packet transfer mode` (1) and sends the `RA Update Request` message (2), the SGSN receives this message (3) with the CGI added by the new BSC. Then the GMM of the SGSN will send a FLUSH-LL message to the old BSC to delete the LLC PDUs. In contrast to the RA update intra-BSC, packet diversion can not be performed, since the new cell is not reachable any more by the old BSC holding the LLC PDUs. The old BSC responds with a FLUSH-ACK message confirming the deletion of these LLC PDUs. Additionally the GMM SGSN must send an `RA Update Accept` message to the new cell in which the MS is located and decide whether to retransmit immediately the LLC-PDUs which were deleted in the old BSC or to rely on the LLC retransmission mechanism after timer expiry if LLC is in the acknowledged mode.

4.2.3.4 Inter-SGSN RA Update

While in the intra-SGSN RA update procedures described in the previous sections, micro-mobility management is performed only regarding cells and RAs, the inter-

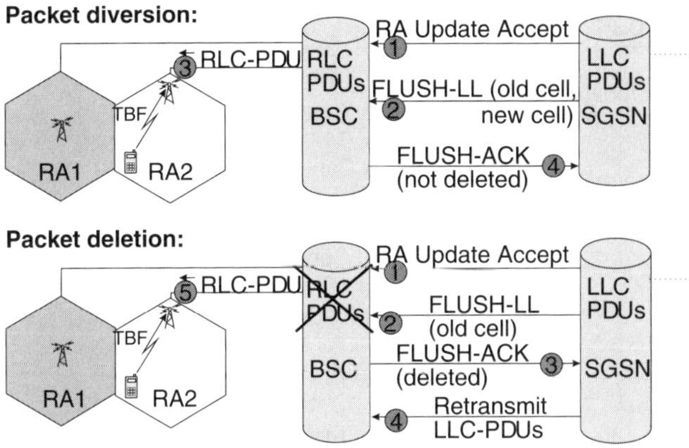

Figure 4.17: RA update Intra-BSC: downlink procedure

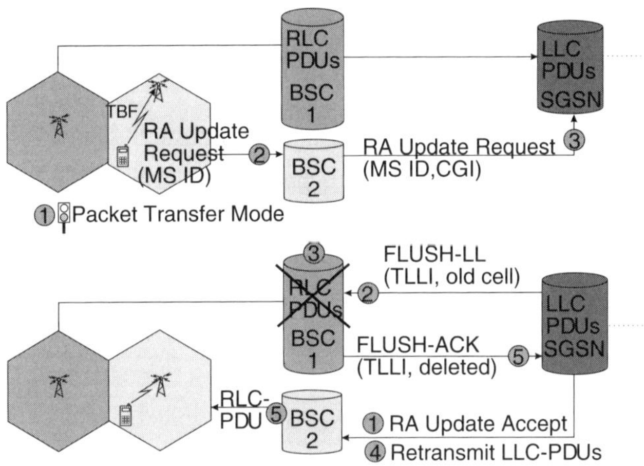

Figure 4.18: RA update Inter-BSC: uplink and downlink procedure

SGSN RA update procedures also involve the GGSNs and the GPRS registers HLR and VLR.

After the MS has changed from an old cell to a new cell that are served by different SGSNs, the MS sends an `RA Update Request` to the new SGSN including the information of the old RA and old P-TMSI signature (see Figure 4.19). The BSS adds the identity of the new cell before passing the message to the SGSN.

After the authentication procedures in the new SGSN, the GTP-C entity (see Section 4.1) of the new SGSN sends an `SGSN Context Request` to the GTP-C entity of the old SGSN to get the MM and PDP contexts for the MS. The old SGSN validates the old P-TMSI signature and responds with an appropriate error cause if it does not match the value stored in the old SGSN. If the old P-TMSI signature was valid or if the new SGSN indicates that it has authenticated the MS, the old SGSN responds with an `SGSN Context Response` message. If the MS is not known in the old SGSN, the old SGSN responds with an appropriate error cause. The old SGSN stores the new

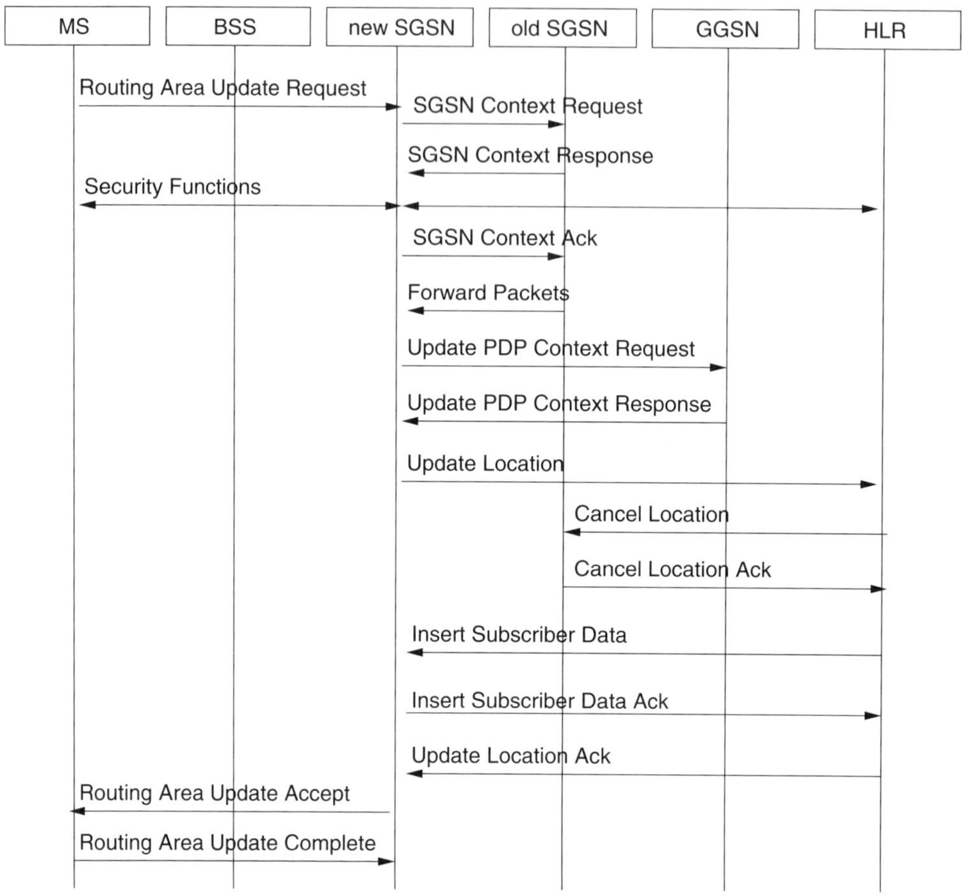

Figure 4.19: Inter-SGSN RA update

SGSN address, to allow the old SGSN to forward data packets to the new SGSN. The old SGSN starts a timer and stops the downlink transfer. Next, security functions concerning the ciphering function are executed.

The new SGSN GTP-C entity sends an `SGSN Context Acknowledge` message to the old SGSN. This informs the old SGSN that the new SGSN is ready to receive data packets belonging to the activated PDP contexts. The old SGSN GTP-C marks in its context that the MSC/VLR association and the information in the GGSNs and the HLR are invalid. This triggers the update of MSC/VLR, the GGSNs, and the HLR, if the MS initiates an RA update procedure back to the old SGSN before completing the ongoing RA update procedure. If the security functions do not authenticate the MS correctly, then the RA update is rejected, and the new SGSN sends a reject indication to the old SGSN. Otherwise the old SGSN forwards the buffered packet data to the new SGSN. Additional packet data received by the old SGSN from the GGSN are also forwarded to the new SGSN. Packet data that had already been sent to the MS in acknowledged mode and that are not yet acknowledged by the MS are forwarded together with the LLC frames that contain the last segment of the user data.

Next, the GTP-C protocol entity (see Section 3.1) of the new SGSN sends an `Update PDP Context Request` message to the GGSN concerned. The GTP *Tunnel Identifiers* (TIDs) are updated at the GGSN so that a new GTP-U tunnel is established between GGSN and the new SGSN. The GGSNs update their PDP context fields and return an `Update PDP Context Response` message. The new SGSN informs the HLR of the SGSN change by sending a MAP (Section 4.1) `Update Location` message to the HLR. The HLR MAP entity sends a `Cancel Location` message to the old SGSN. The old SGSN removes the MM and PDP contexts, after a timeout. The timer allows the old SGSN to complete the forwarding of packet data. It also ensures that the MM and PDP contexts are kept in the old SGSN in case the MS initiates another inter-SGSN RA update before completing the ongoing RA update to the new SGSN. The MAP entity of the old SGSN acknowledges with a `Cancel Location Acknowledge` message.

The HLR sends an `Insert Subscriber Data` message to the new SGSN. The new SGSN validates the MS's presence in the new RA. If due to regional subscription restrictions the MS is not allowed to be attached in the RA, the SGSN rejects the `RA Update Request` with an appropriate cause and may return an `Insert Subscriber Data Acknowledge` message to the HLR. If the authentication is successful, the SGSN constructs an MM context for the MS and returns an `Insert Subscriber Data Acknowledge` message to the HLR. The HLR acknowledges the `Update Location` message by sending an `Update Location Acknowledge` message to the new SGSN. The GMM entity at the new SGSN responds to the MS with an `RA Update Accept` message. The MS acknowledges the new P-TMSI with an `RA Update Complete` message. If the `RA Update Complete` message confirms the reception of LLC PDUs that were additionally forwarded from the old SGSN, then these LLC PDUs are discarded by the new SGSN.

4.2.4 GPRS Roaming

Customers of emerging GPRS services will want to be able to use their terminal globally as they are used to in GSM networks. To be able to offer GPRS roaming, operators worldwide need an economic, secure and reliable platform to interconnect their networks. A solution is a *GPRS Roaming Exchange* (GRX) service offered by service providers that operate a global GRX network. Proposals of such a network [36] are based on an IP *Virtual Private Network* (VPN) platform as shown in Figure 4.20. A VPN has the benefit of being able to share a common underlying infrastructure with other IP services whilst providing a closed and secure IP network dedicated to the GPRS roaming service. A number of *Points of Presence* (POP) are available for the connection of domestic GPRS networks. The *Customer Edge Router* provides the border gateway functionality for the GPRS operator's network. The *Provider Edge Router* acts as the GRX border gateway router at the edge of the GRX network. With such a concept it is also possible to interconnect several GRX networks of multiple GRX providers.

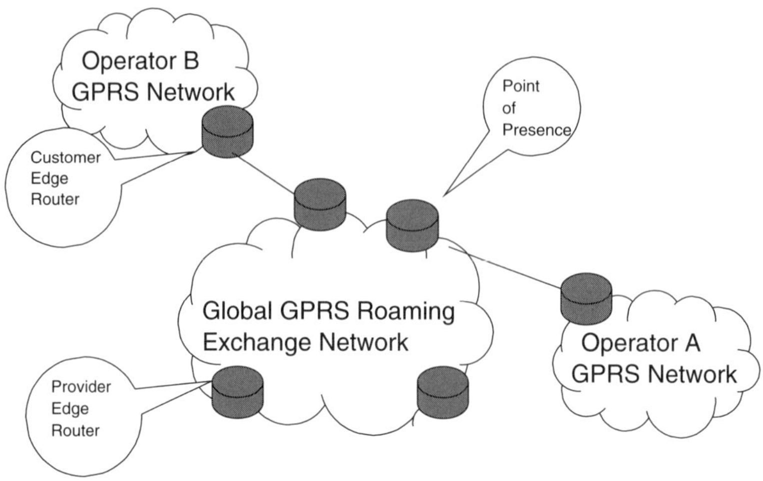

Figure 4.20: GRX network with two operators connected

4.3 Session Management

Session Management (SM) in GPRS comprises all signaling functions necessary for the access to external *Packet Data Networks* (PDNs). For interworking with external IP networks the IP interworking model is defined in [10].

4.3.1 The IP Interworking Model

Network interworking is required whenever a PLMN is involved in communications with another network to provide end-to-end transport. The PLMN interconnects in a manner consistent with that of a *Packet Data Network* (PDN) defined by the requirements for interworking with *Public Switched Data Networks* (PSDNs) X.75 [72].

GPRS supports interworking with networks based on the Internet Protocol. These interworked networks may be either intranets or the Internet [10]. When interworking with IP networks, GPRS can operate IPv4 or IPv6. The interworking point with IP networks is at the G_i reference point as shown in Figure 4.21. The GGSN is the access point of the GPRS network. In this case the GPRS network will look like any other IP network or subnetwork.

The G_i reference point is between the GGSN and the external IP network. From the external IP network's point of view, the GGSN is seen as a normal IP router. The access to the Internet, an intranet or *Internet Service Provider* (ISP) may involve specific functions such as user authentication, user's authorization, end-to-end encryption between MS and Intranet/ISP and the allocation of a dynamic address belonging to the PLMN/Intranet/ISP addressing space. For this purpose the GPRS PLMN may offer:

- either direct transparent access to the Internet
- or a non-transparent access to an intranet/ISP.

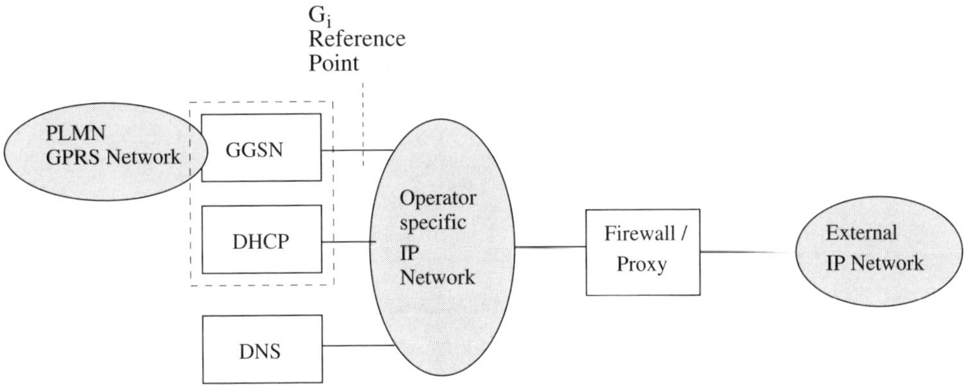

Figure 4.21: IP interworking

In the transparent case the mobile station is given an address belonging to the operator's addressing space (see Figure 4.21). The address is a public IP address given either at subscription in which case it is a static address or at PDP context activation in which case it is a dynamic address. This address is used for packet forwarding between the Internet and the GGSN and within the GGSN to tunnel user data through the core network.

The transparent case provides at least a basic ISP service. As a consequence of this it may therefore provide a bearer service for a tunnel to a private Intranet. In the non-transparent case the mobile station is given an address belonging to the Intranet/ISP private addressing space, static or dynamic. Packet forwarding within the GGSN and on the Intranet/ISP requires a link between the GGSN and an address allocation server, like *Remote Authentication Dial In User Service* (RADIUS) and *Dynamic Host Configuration Protocol* (DHCP) belonging to the Intranet/ISP. This server responses to authentication requests at a PDP context activation.

4.3.2 PDP Context Handling

The main function of *Session Management* (SM) [11] is to support PDP context handling of the MS. The SM comprises procedures for

- identified PDP context activation, deactivation and modification
- anonymous PDP context activation and deactivation.

Anonymous access does not need a GMM context to be established, and there is no P-TMSI assigned to the MS. In the following, only identified access will be regarded.

QoS profile negotiation between MS and the network is performed within the procedures for the activation or modification of PDP contexts. Therefore, SM procedures play an important role in the GPRS mechanisms for QoS support. The states and procedures of SM both on MS and on the network side will thus be explained in the following sections.

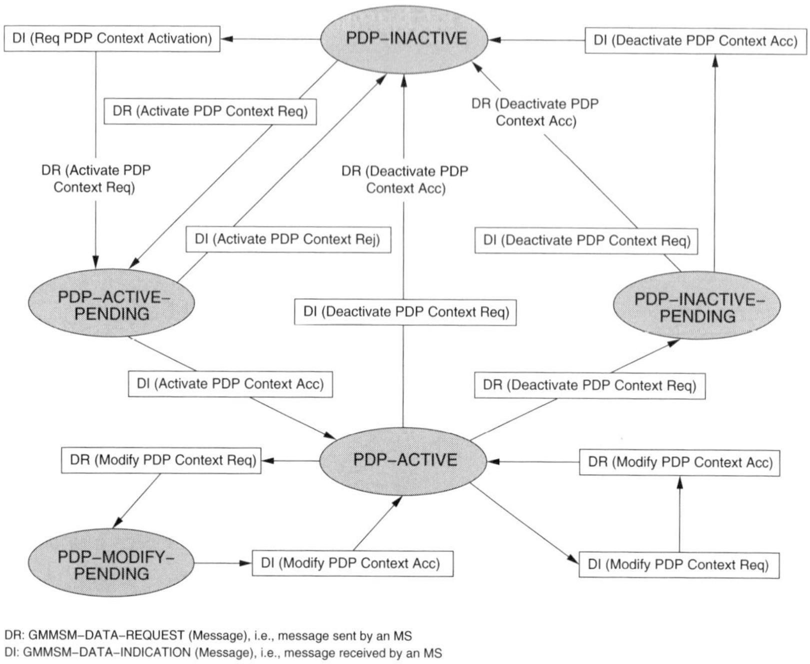

Figure 4.22: SM states in the MS

4.3.3 Session Management States

This section deals with the SM states defined for both the MS and the network. SM states are described for one SM entity, each SM entity being associated with one PDP context [3].

Figure 4.22 provides an overview of the possible states of an SM entity in the MS. There are five SM states possible:

PDP-INACTIVE: There is no PDP context existing.

PDP-ACTIVE-PENDING: A PDP context activation was requested by the MS.

PDP-INACTIVE-PENDING: The MS has requested the deactivation of its PDP contexts.

PDP-ACTIVE: The PDP context is active.

PDP-MODIFY-PENDING: A modification of the PDP context was requested by the MS.

The state PDP-MODIFY-PENDING is only valid for GPRS Release 99-capable MSs. Release 97/98 does not provide the possibility of PDP context modification initiated by MSs.

As illustrated in Figure 4.23, there are five SM states on the network side:

PDP-INACTIVE: The PDP context is not active.

PDP-ACTIVE-PENDING: A PDP context activation was initiated by the network.

PDP-INACTIVE-PENDING: The network has requested the deactivation of the PDP context.

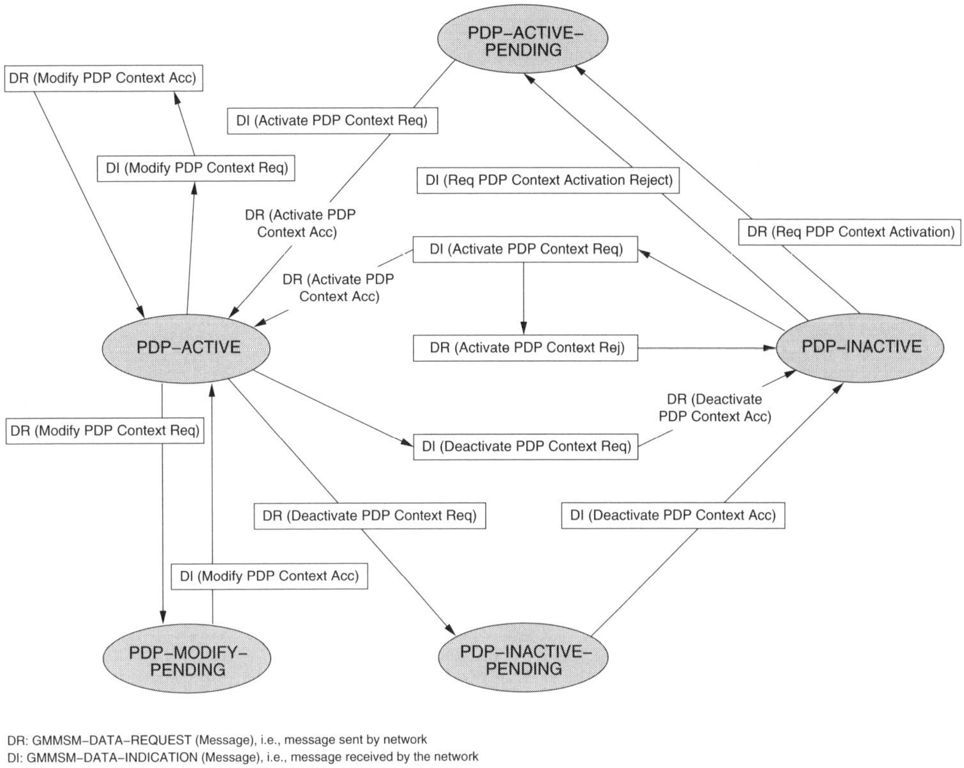

Figure 4.23: SM states on the network side

`PDP-ACTIVE`: The PDP context is active.

`PDP-MODIFY-PENDING`: A modification of the PDP context has been requested by the network.

Reception of a `Modify PDP Context Request` message and the associated state transitions are not possible for GPRS Release 97/98.

4.3.4 Session Management Procedures

In the following subsections PDP context activation, modification, and deactivation procedures will be explained in brief.

The purpose of the PDP context activation procedure is to establish a PDP context between the MS and the network for a specific QoS on a specific N-SAPI (see Section 3.3). The PDP context activation may be initiated by the MS or the initiation may be requested by the network.

4.3.4.1 PDP Context and Traffic Flow Template

In GPRS Release 97/98, there is only one PDP context possible per MS. In GPRS Release 99, each PDP address may be described by one or more PDP contexts in the MS or the network. The first PDP context activated for a PDP address is called

Octet \ Bit	8	7	6	5	4	3	2	1
1	Type = 130 (Decimal)							
2–3	Length							
4	Res–erved	AA	Res–erved	Order	NSAPI			
5	X	X	X	X	SAPI			
6–8	QoS Sub							
9–11	QoS Req							
12–14	QoS Neg							
15–16	Sequence Number Down (SND)							
17–18	Sequence Number Up (SNU)							
19	Send N–PDU Number							
20	Receive N–PDU Number							
21–22	Uplink Flow Label Signalling							
23	Spare 1 1 1 1				PDP Type Organization			
24	PDP Type Number							
25	PDP Address Length							
26–m	PDP Address [1..63]							
m+1	GGSN Address for signalling Length							
(m+2)–n	GGSN Address for signalling [4..16]							
n+1	APN length							
(n+2)–o	APN							
o+1	Spare (sent as 0 0 0 0)				Transaction Identifier			

Figure 4.24: PDP context information element

the *primary PDP context*, whereas all additional contexts associated with the same PDP address are called *secondary PDP contexts.* When more than one PDP context is associated with a PDP address, there will be a *Traffic Flow Template* (TFT) for each additional context. The TFT will be sent transparently via the SGSN to the GGSN to enable packet classification and policing for downlink data transfer [29]. A TFT consists of from one up to eight packet filters, each identified by a unique packet filter identifier. A packet filter has an evaluation precedence index that is unique within all TFTs associated with the PDP contexts that share the same PDP address. The MS manages packet filter identifiers and their evaluation precedence indexes, and creates the packet filter contents. The packet filter specifies attributes for incoming and outgoing IP packets. The IP datagram attributes have to match at least for the packet filter of one evaluation precedence index of the TFT to pass the GGSN.

4.3.4.2 PDP Context Information Element

All PDP context information is stored within an appropriate *Information Element* (IE) [9]. In Figure 4.24 the fields corresponding to QoS Subscribed (QoS Sub), QoS

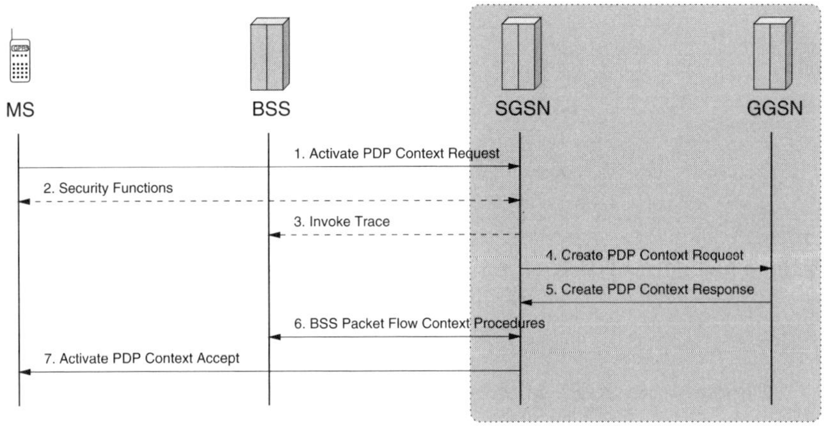

Figure 4.25: PDP context activation procedure

Requested (QoS Req), and QoS Negotiated (QoS Neg), as well as the PDP Address, are emphasized. They play an important role in the implementation of the mechanisms to support QoS.

All QoS-related information, whether subscribed, requested, or negotiated, is included in a QoS profile. The associated IE is described in Section 4.4.4.

4.3.4.3 PDP Context Activation, Modification and Deactivation

Figure 4.25 outlines the message sequence for a successful MS-initiated PDP context activation procedure. Upon receiving an `Activate PDP Context Request` message the SGSN will initiate procedures to set up PDP contexts. This includes, first of all, subscription checking. Security functions may be executed as well.

Furthermore, it can be checked whether the SGSN load is below a predefined threshold and if the related radio cell is able to support the requested QoS. The current cell status is based on flow control information from the BSS sent over BSSGP (see Section 3.2).

If the serving host is outside the GPRS core network, i.e. the GGSN is involved in a traffic flow between an MS and a serving host, a `Create PDP Context Request` is sent to the GGSN. The negotiated QoS profile from the SGSN must also meet possible restrictions due to the GGSN's load and capabilities. The GGSN then creates a new entry in its PDP context table to be able to route PDP PDUs between the SGSN and the external PDN. If there is already a primary PDP context active for the MS, the secondary PDP contexts are provided with a TFT associated with the relevant PDP address. If these GGSN procedures have been successful, the GGSN sends a `Create PDP Context Response` back to the SGSN, containing the possibly modified negotiated QoS profile.

If the requested QoS has been accepted by both SGSN and GGSN, it is returned to the MS as the negotiated QoS profile by an `Activate PDP Context Accept` message, and to the BSS in charge during BSS PFC procedures (see Section 3.2).

Otherwise, if the cell is congested or the SGSN or GGSN load thresholds are exceeded, either the `PDP Context Activation Request` is rejected with a `PDP Context Activation Reject` message, or the SGSN or GGSN attempt to negotiate a QoS pro-

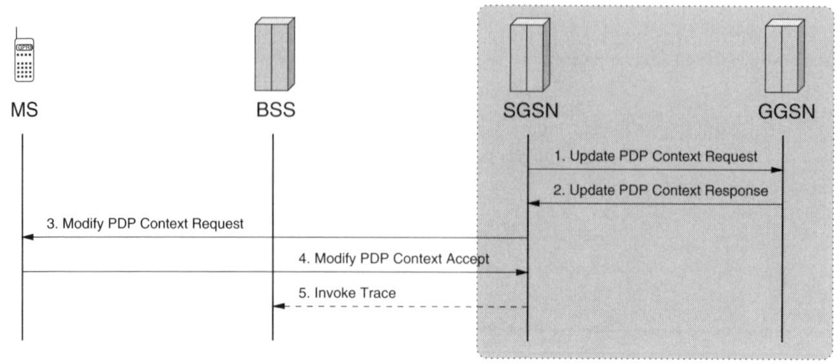

Figure 4.26: SGSN-initiated PDP context modification procedure

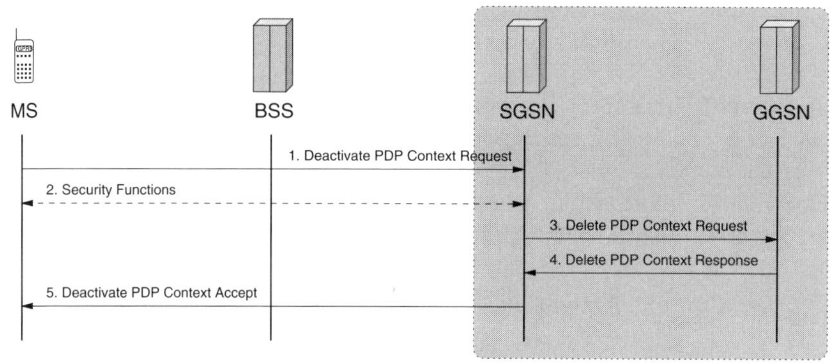

Figure 4.27: PDP context deactivation procedure

file lower than requested. If the BSS cannot support the requested QoS the SGSN suggests best effort, which either the MS accepts, or it deactivates the PDP context.

While in GPRS Release 97/98 only the SGSN may initiate this procedure, PDP context modification may be initiated by MS, SGSN, or GGSN in a Release 99 conforming network. It takes place, e.g., in case of inter-SGSN RA updates, when the new SGSN is unable to maintain the QoS negotiated, or due to changed load conditions in the radio cell.

Figure 4.26 shows an SGSN-initiated PDP context modification procedure. The GGSN is informed about the context to modify by an `Update PDP Context Request` message, which it answers with an `Update PDP Context Response`. The MS concerned either accepts the following `Modify PDP Context Request`, or initiates a PDP context deactivation procedure.

Deactivation of a PDP context may also be initiated by MS, SGSN, or GGSN. Figure 4.27 shows the message sequence for an MS-initiated PDP context deactivation procedure. As on PDP context activation, security functions may first be conducted when an SGSN receives a `Deactivate PDP Context Request`. Thereafter, a `Delete PDP Context Request` is sent to the GGSN, which is answered by a `Delete PDP Context Response`. When both SGSN and GGSN have deleted the according PDP context, the MS is informed of completion by a `Delete PDP Context Accept` message.

The MS may include the *Tear Down Indicator Information Element* in the `Deactivate PDP Context Request` in order to indicate that all active PDP contexts sharing the same PDP address as the PDP context associated with this specific *Transaction Identifier* (TI) will be deactivated. If the tear down indicator information element is not included, only the PDP context associated with this specific TI is deactivated.

4.4 Quality of Service Management

4.4.1 General Aspects

To elaborate a concept for a traffic management functionality based on application QoS requirements, it is necessary to define the implications of the term QoS itself.

In this context, QoS shall be defined as "the collective effect of service performance, which determines the degree of satisfaction of a user of the service" [77].

It is composed of numerous parameters. In the following the most important ones will be described.

4.4.1.1 QoS Characteristics

Technically, QoS refers to an aggregation of system performance measures. The five most important of these are [77, 49, 34]:

Availability: The availability of a network, its components, or even a service, should ideally approximate 100%. Even a value like 99.8% means about an hour and a half of down time per month. Thus, consumer satisfaction largely depends on this parameter.

Throughput: This is the effective data transfer rate available to an application, measured in bit/s. It depends on – but is explicitly not the same as – the maximum capacity, or bandwidth, of the network. Multiple connections sharing a transmission link, packet errors and losses during transfer, overhead imposed by protocol headers as well as characteristics of the nodes on the transmission path, such as buffer capacity or processing power, lower the throughput at disposal for an application.

Packet loss: Network elements, like switches and routers, are equipped with buffered queues to adopt to link congestion to some extent. However, if a link remains congested for too long, this will result in a buffer overflow and thus a loss of data. In a mobile radio network packets may additionally get lost owing to the special conditions on the radio interface. Both cases usually result in a retransmission of the packet increasing the total transmission time.

Latency: Latency or delay is understood as the time taken by data to travel from its source to its destination. Thus, it may also be referred to as end-to-end delay, and is an important aspect of the perceived QoS. Since long delays reduce the interactivity of communication, interactive real-time (RT) applications are especially affected by it, while non-interactive RT applications show more sensitivity to a variation in delays, also called jitter. Non-real-time (NRT) applications are usually not delay-sensitive.

Table 4.1: Varied sensitivities of network traffic types

Traffic type	Sensitivities			
	Bandwidth	Loss	Delay	Jitter
Voice	Very low	Medium	High	High
E-commerce	Low	High	High	Low
Transactions	Low	High	High	Low
E-mail	Low	High	Low	Low
Telnet	Low	High	Medium	Low
Casual browsing	Low	Medium	Medium	Low
Serious browsing	Medium	High	High	Low
File transfers	High	Medium	Low	Low
Video conferencing	High	Medium	High	High
Multicasting	High	High	High	High

Various components add up to the end-to-end delay of a packet on a transmission path:

- *Transmission delay:* the time it takes to put all bits of a packet onto the link
- *Propagation delay:* the time it takes for a bit to traverse a link (e.g., at the speed of light)
- *Processing delay:* the time it takes to process a packet in a network element (e.g., routing it to the output port)
- *Queuing delay:* the time a packet must wait in a queue before it is scheduled for transmission

In a mobile radio environment there may be additional delays adding to the delays mentioned, caused by random access (RA) or paging mechanisms.

Jitter: Jitter, or latency variation, may be induced by various causes, e.g., variations in queue length, variations in the processing time needed to reorder packets that arrived out of order because they traveled over different paths, and variations in the processing time needed for re-assembly of packets segmented before being transmitted. Again, interactive RT applications, especially, are sensitive to delay jitter, as well as non-interactive RT applications. The latter may be able to adjust their playback point, i.e. the time offset between playback of consecutive packets, based on changes in the jitter value, and are then called "adaptive" applications. Packets arriving after their playback point has passed are generally not useful to the application, and are, in most cases, discarded.

As mentioned above, applications vary significantly in their QoS requirements (see Table 4.1, [49]). Interactive RT applications like VoIP have the most stringent demands on system performance, especially concerning delay and delay jitter. While non-interactive RT applications, like streaming audio or video, largely depend on small jitter values and, to a certain extent, on packet delay, NRT applications, like FTP or e-mail, are in most cases delay and jitter independent, but need throughput values as good as possible. Compared to delay or jitter, occasional loss of packets does not have a strong impact on the performance of RT applications of any kind, and solely reduces the throughput, due to packet retransmissions, when regarding NRT applications. Network availability should be as high as possible, but is not a parameter under the influence of mechanisms for QoS provisioning. Thus, such mechanisms have to primarily concentrate on optimizing delay and throughput measures on behalf of application requirements.

4.4.1.2 QoS Provisioning

There are two basic mechanisms for providing adequate QoS based on delay and throughput measures:

- Plentiful capacity
- Traffic management

With *plentiful capacity*, the assumption is that there is enough capacity available in the network so that no explicit mechanisms have to be provided to ensure QoS. This implies that there are enough:

- High-capacity links
- Fast processors
- Plentiful buffers

This is perhaps a reasonable assumption for a controlled, localized environment, such as a corporate LAN. It is unlikely that such assumptions will hold true across a global network such as the Internet. While the cost of bandwidth, memory, and processing are coming down (MOORE'S law), there is still a high cost to be paid for the high-end equipment assumed by this model. Even today, the cost of high-capacity Wide Area Network (WAN) links is quite high, as are the costs of high-speed routers and switches.

A variation of the *plentiful capacity* model is that, even if not enough bandwidth is not available, applications can adapt to varying bearer quality. Adaptive applications certainly exist, but they are restricted to a certain range of parameters to provide useful service. Therefore, beyond a certain point, it may not be possible to provide adequate QoS without any explicit controls.

The second model of provisioning QoS is *traffic management*. The fundamental idea here is that traffic can be differentiated and classified into different levels of service. The granularity of differentiation may be a small set of classes (e.g., simple priority) or could be as fine as each application flow. Some control must be exerted on how much traffic of each class is allowed into the network, based on the available resources – this may be done statically (provisioning) or dynamically (signaling for resource reservations). Additionally, network elements must manage the processing and queueing of packets in such a way that appropriate, differentiated services are provided to the packets.

There are two sub-categories of traffic management:

Reservation-based: In this model, resources for traffic are explicitly identified and reserved. Network nodes classify incoming packets and use the reservations to provide differentiated services. Typically, a dynamic resource reservation setup protocol is used, in conjunction with admission control, to set up reservations. Further, the nodes use intelligent processing, e.g., *Random Early Detection* (RED), and queuing mechanisms, e.g., *Weighted Fair Queueing* (WFQ), to service packets.

Reservation-less: In this model, no resources are explicitly reserved. Instead, traffic is differentiated into a set of classes, and network nodes provide priority-based treatment of these classes. It may still be necessary to control the amount of traffic in a given class that is allowed to be injected into the network, to preserve the QoS being provided to other packets of the same class.

The GPRS specification follows the second approach, defining a set of QoS parameters that are combined to QoS profiles designed to meet the requirements of one kind of traffic class each. Connection Admission Control can be performed to ensure that the QoS negotiated for the packet data flows already in the system remains undiminished, and that any kind of traffic will be served at least to a certain degree.

4.4.2 GPRS QoS Management

The performance of different applications experienced by a user is influenced by all network elements located on the path between the client and the server. Depending on the location of the server, also called the host, different end-to-end QoS scenarios have to be regarded.

In Figure 4.28, a typical GPRS network (PLMN 1) is shown, with its ingress/egress nodes and neighboring external networks that may lie in the transmission path of data

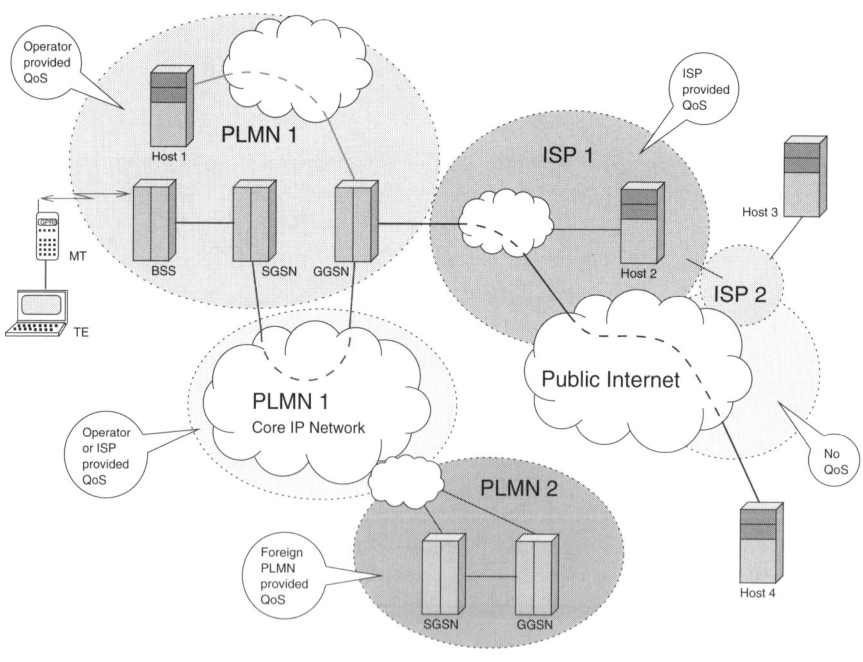

Figure 4.28: End-to-end QoS for different server locations

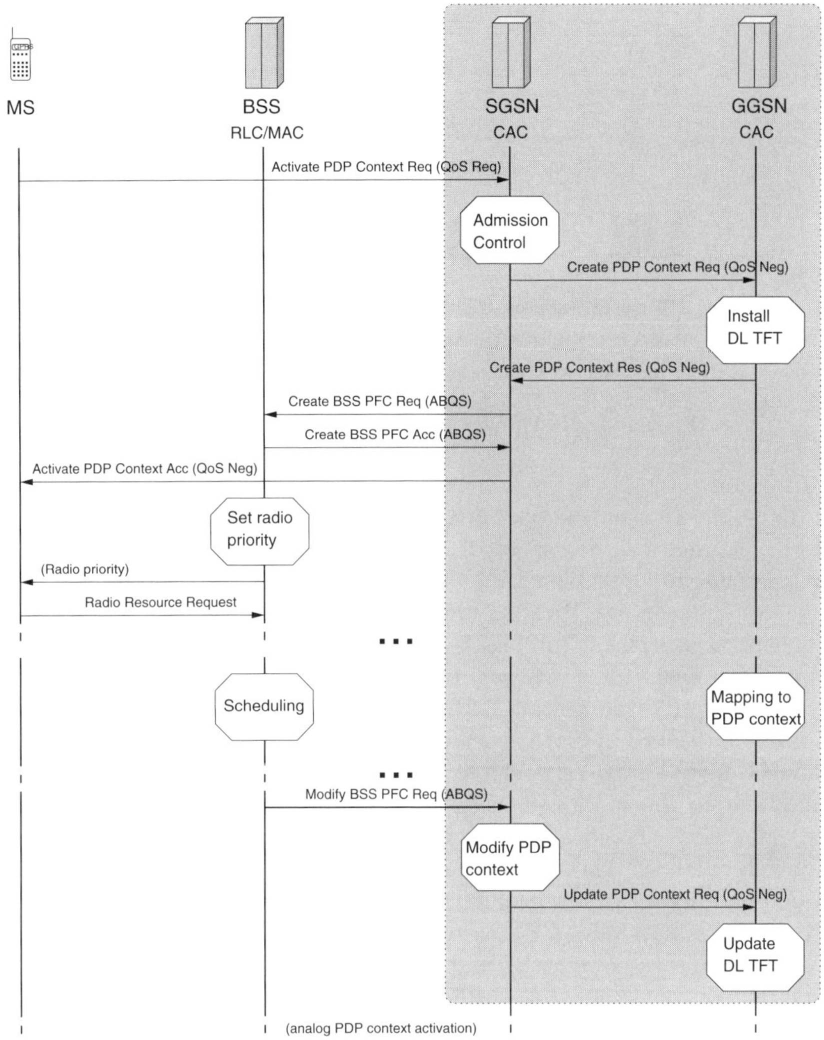

Figure 4.29: QoS negotiation and renegotiation procedures (example)

packets requested by a GPRS terminal TE. Only in the case of the requested service being offered by Host 1 full control of all nodes and links located in the transmission path is given to the GPRS network operator. Even then, it might be necessary for the operator to rent transport resources for his core IP network from a separate network operator (PLMN 1 Core IP Network) and partially lose the QoS control. Any other host (Host 2–Host 3) being the GPRS subscriber's requested target host, withdraws a substantial part of the intermediate network from the GPRS operator's controlled domain.

To be in a position to offer QoS to the subscriber, the operator needs to have contracts with the owners of the external networks, so-called *Service Level Agreements* (SLAs), that assure an appropriate QoS mapping between the separate domains.

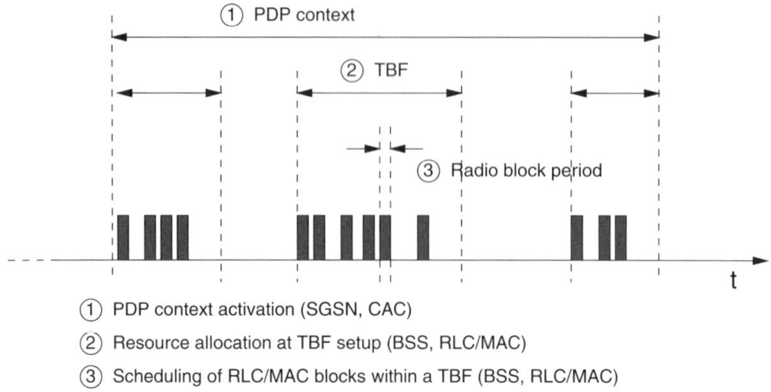

Figure 4.30: Three-stage model of QoS management

Should the public Internet be part of the transmission path (Host 4) no more than best-effort treatment can be expected.

As seen in Chapter 3, Section 2.2.2 and Section 4.3, there are several logical entities and protocols involved in GPRS QoS management.

PDP contexts as well as QoS profiles are negotiated between MS and SGSN. The BSS is provided with a *Packet Flow Context* (PFC) containing the *Aggregate BSS QoS Profile* (ABQP) (see Section 3.2) and is responsible for resource allocation on a TBF base and scheduling of packet data traffic with respect to the relevant QoS profiles negotiated. Moreover, it has to regularly inform the SGSN about the current load conditions in the radio cell. The tasks of the GGSN comprise mapping PDP addresses as well as classification of incoming traffic from external networks on behalf of downlink TFTs. The *GPRS Registers* (GRs) hold the QoS-related subscriber information and deliver it, on demand, to the GSNs.

In Figure 4.29, part of a GPRS session is schematically outlined, depicting the instances involved, messages exchanged, and parameters negotiated for PDP context, PFC and TFT installation and renegotiation.

From a time-scale point of view, the mechanisms for QoS management within the GPRS might be regarded as a three-stage model (see Figure 4.30). On PDP context activation, the QoS parameters are negotiated. As long as the PDP context remains active, these parameters should be guaranteed, unless there is a QoS renegotiation. The QoS profile is considered both for each TBF and for each radio block period. At TBF setup, radio resources, like the number of PDCHs, the associated USFs and TFIs, are assigned according to the negotiated QoS parameters. During the TBF, radio blocks are scheduled at the BSS in competition with other existing TBFs in the radio cell. This scheduling function has to be done considering the QoS profiles of the PDP contexts associated with the TBFs.

A QoS profile can be considered as a single parameter value that is defined by a unique combination of attributes. There are numerous QoS profiles available based on permutations of these different attributes, but each mobile network operator must choose to support only a limited subset, reflecting their planned range of GPRS subscriptions. In the following subsections, the QoS attributes defined in GPRS Release 97/98, as well as the changes made for Release 99, will be explained.

4.4.3 QoS Attributes According to GPRS Release 97/98

A QoS profile defines the QoS within the range of the following service classes [25, 29].

4.4.3.1 Precedence classes

Under normal circumstances the network should try to meet all profiles' QoS agreements. The precedence specifies the relative importance to keep the conditions even under critical circumstances, e.g., momentarily high network load. The various precedence classes are presented in Table 4.2.

4.4.3.2 Delay classes

The packet delay is defined by the time needed for transmission from one GPRS SAP to another. Delays outside the system, e.g., in transit networks, are not considered. [25] determines four delay classes (see Table 4.3). The network operator has to provide for convenient resources on the air interface to be able to serve the number of participants with a certain delay class expected within each cell. Although there is no need for all delay classes to be available, at least *best effort* has to be offered.

4.4.3.3 Reliability classes

Data services generally require a low residual *Bit Error Ratio* (BER). Erroneous data is usually useless, while incorrectly received speech only leads to a worse perception. Reliability of data transmission is defined within the scope of the following cases:

- probability of loss of data
- probability of out-of-sequence data delivery
- probability of multiple delivery of data and
- probability of erroneous data.

Table 4.2: Precedence classes

Precedence class	Identifier	To be served
1	High priority	preferably before classes 2 and 3
2	Normal priority	preferably before class 3
3	Low priority	without preference

Table 4.3: Delay classes

Delay class	128 byte packet		1 024 byte packet	
	Mean delay [s]	95 % [s]	Mean delay [s]	95 % [s]
1 (predictive)	0.5	1.5	2	7
2 (predictive)	5	25	15	75
3 (predictive)	50	250	75	375
4 (best effort)	unspecified			

Table 4.4: Reliability class

Reliability classes	GTP mode	LLC frame mode	LLC data mode	RLC block mode	Traffic type security
1	ACK	ACK	PR	ACK	NRT traffic, error sensitive, loss sensitive
2	UACK	ACK	PR	ACK	NRT traffic, error sensitive, slightly loss sensitive
3	UACK	UACK	PR	ACK	NRT traffic, error sensitive, not loss sensitive
4	UACK	UACK	PR	UACK	RT traffic, error sensitive, not loss sensitive
5	UACK	UACK	UPR	UACK	RT traffic not error sensitive, not loss sensitive
(U)ACK	(Un)acknowledged			NRT	Non-Realtime
PR/UPR	Protected/Unprotected			RT	Realtime

The reliability classes specify the requirements for each layer's services. The combination of different modes of operation of the GPRS specific protocols GTP, LLC, and RLC, explained in Chapter 3, support the reliability requirements of various applications, e.g., *Real-Time* (RT) or *Non Real-Time* (NRT). The reliability classes are summarized in Table 4.4.

4.4.3.4 Peak throughput classes

User data throughput is specified within the scope of a set of throughput classes that characterize the expected bandwidth for a requested PDP context. It is defined by the choice of peak and mean throughput class. Peak throughput is measured in byte/s at the reference points G_i and R (see Figure 2.12). Peak throughput specifies the maximum rate at which data is transmitted within a certain PDP context. No guarantee is given that this data rate is actually achieved at any time during transmission. Rather, this depends on the resources available and the capabilities of the MS. The operator may limit the user data rate to the peak data rate agreed on, even if there is capacity left for disposal. The peak throughput classes are presented in Table 4.5.

4.4.3.5 Mean throughput classes

Like peak throughput, mean throughput is also measured in byte/s at the reference points G_i and R. It specifies the average rate data transmitted within the time remaining for a certain PDP context. The operator may limit the user data rate to the mean data rate negotiated, even if excessive capacity is available. If *best effort* has been agreed on as the throughput class, throughput is made available to a MS whenever there are resources needed and at its disposal. Table 4.6 summarizes the classes of mean throughput.

Table 4.5: Peak throughput classes

Peak throughput class	Peak throughput	
	[byte/s]	[kbit/s]
1	up to 1 000	8
2	up to 2 000	16
3	up to 4 000	32
4	up to 8 000	64
5	up to 16 000	128
6	up to 32 000	256
7	up to 64 000	512
8	up to 128 000	1 024
9	up to 256 000	2 048

Table 4.6: Mean throughput classes

Mean throughput class	Mean throughput	
	[byte/h]	≈ [bit/s]
1	100	0.22
2	200	0.44
3	500	1.11
4	1 000	2.2
5	2 000	4.4
6	5 000	11.1
7	10 000	22
8	20 000	44
9	50 000	111
10	100 000	220
11	200 000	440
12	500 000	1 110
13	1 000 000	2 200
14	2 000 000	4 400
15	5 000 000	11 100
16	10 000 000	22 000
17	20 000 000	44 000
18	50 000 000	111 000
31	Best effort	

4.4.4 QoS Profile Information Element

All QoS-related information to be exchanged between MS and SGSN is stored in a QoS profile. The appropriate IE is shown in Figure 4.31. It consists of an *Information Element Identifier* (IEI), a length field, five fields that contain the values of the service classes (see Section 4.4.3), and three fields filled with spare bits [11]. IEI and length fields are not part of a PDP context IE (see Section 4.3.4.2).

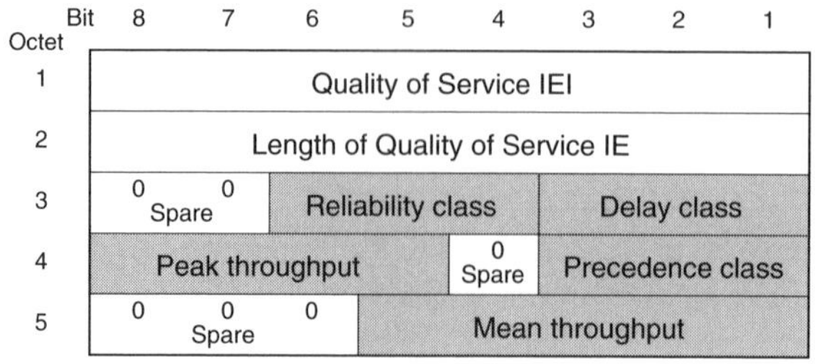

Figure 4.31: QoS profile information element

4.4.5 QoS in GPRS Release 99

It is evident from the above description that the QoS architecture defined in GPRS Release 97/98 shows some major drawbacks (see also [62, 107]):

1. The BSS is not aware of the negotiated QoS profile. This restricts the ability of the BSS to perform scheduling and resource management on the radio interface.
2. Neither MS nor GGSN can influence the QoS profile, even if they detect congestion in external networks as well as changing radio conditions or varying application requirements.
3. It is only possible to have one QoS profile for every PDP context that is associated with a specific service access. Thus, there can only be one QoS profile utilized by all applications for one PDP address.

GPRS Release 99 defines several further QoS parameters with finer grained properties to meet requirements on different levels of service for applications [30]:

- maximum bitrate
- guaranteed bitrate
- delivery order
- maximum SDU size
- SDU format information
- SDU error ratio
- residual bit error ratio
- delivery of erroneous SDUs
- transfer delay
- traffic handling priority
- allocation/retention priority
- source statistics descriptor ('speech'/'unknown')

Additionally, there are four distinct *traffic classes* introduced, with different parameters specifying their QoS requirements (see Table 4.7):

- *conversational*
- *streaming*
- *interactive*
- *background*

Delay-sensitive services belonging to the *conversational* class, for example, do need absolute guarantees in terms of *guaranteed bitrate* and *transfer delay* attributes, while for *background* traffic only bit integrity is necessary.

In GPRS Release 97/98, the BSS cannot use QoS profile information to schedule resources on a continuous data flow, neither on the downlink, since there is no

Table 4.7: End-user performance expectations for selected services belonging to different traffic classes [27]

Traffic class	Medium	Application	Data rate (kbit/s)	One-way delay
Conversational	Audio	Telephony	4–25	< 150 ms
	Data	Telnet	< 8	< 250 ms
Streaming	Audio	Streaming (HQ)	32–128	< 10 s
	Video	One-way	32–384	< 10 s
	Data	FTP	—	< 10 s
Interactive	Audio	Voice messaging	4–13	< 1 s
	Data	Web-browsing	—	< 4 s/page

mechanism provided to download the QoS profile from the SGSN, nor on the uplink, because there is no QoS information available from the MS. With the introduction of Release 99, the BSS is not only provided with QoS profiles on a PFC base, but also with the ability to modify the QoS profile associated with a data flow in case of changing load conditions. Likewise, MS and GGSN may initiate QoS profile renegotiation, either because of changing application requirements, or due to congestion or a change in radio link quality.

Release 99 also solves the Release 97/98 problem of having only one PDP context installed per PDP address; thus all different applications running on top of this PDP address with one single QoS profile, independent of their specific requirements on, e.g., delay or reliability. GPRS Release 99 provides the possibility to install multiple PDP contexts per PDP address. Each PDP context is uniquely associated with a TFT which identifies the traffic flow. This makes it possible to assign different QoS profiles to simultaneous traffic flows, so that each application may receive the appropriate QoS requirement.

GPRS Release 99 conformant architectural components are not deployed in all GPRS networks today. For interworking purposes with GPRS Release 97/98 conformant network equipment, mapping rules between the different releases are defined [30]. These mapping rules will be used for application-level QoS profile definitions, *Admission Control* (AC) rules, and application-based scheduling.

4.4.6 QoS in the Core Network Applying DiffServ

Differentiated Services (DiffServ) is an approach to providing QoS within IP networks in a scalable manner. It is a relatively simple and coarse method of providing differentiated classes of service for various application and subscriber requirements in IP networks [35].

The DiffServ architecture (see Figure 4.32) comprises a small, well-defined set of building blocks from which a variety of aggregate behaviors may be built. The QoS support may be end-to-end or intra-domain. Both quantitative performance requirements (e.g., peak throughput) and requirements based on relative performance (e.g. class differentiation) are supported.

In DiffServ, the network allows applications to negotiate one of the several different services through connection admission control (CAC). The packets generated by applications are treated differently by the network. With proper engineering, including boundary policing, DiffServ can provide expedited handling appropriate for a wide

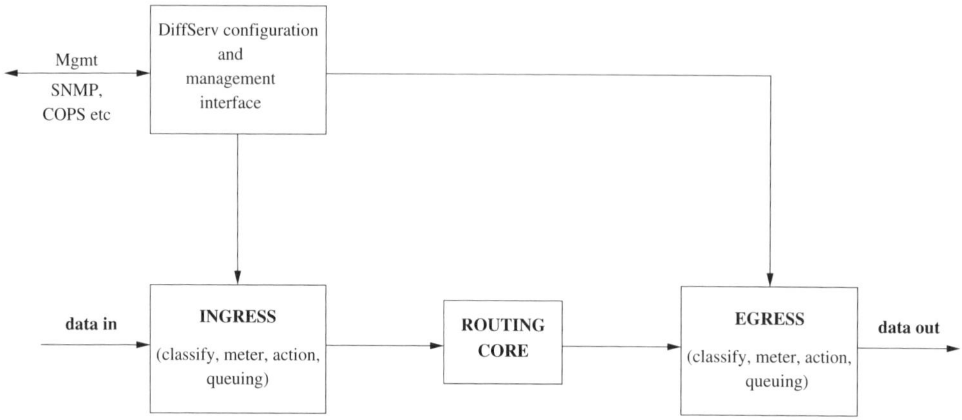

Figure 4.32: Differentiated Services architecture

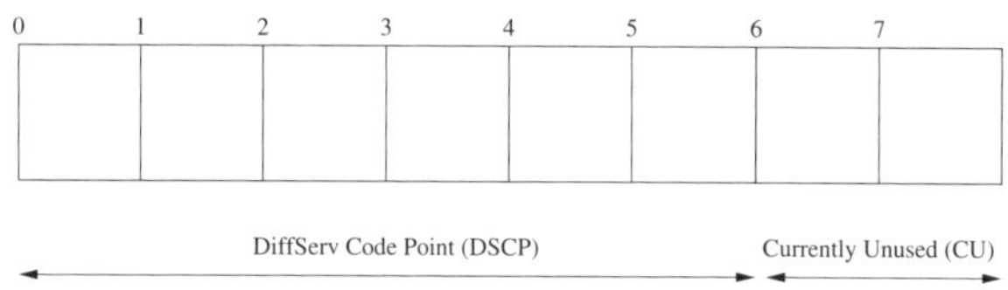

Figure 4.33: IPv4 Type of Service (TOS) octet or IPv6 Traffic Class octet

class of applications, including delay-critical applications. DiffServ-capable routers need to track a small number of per-hop behaviors and they serve packets based on a single byte.

4.4.6.1 DiffServ Field

A bit pattern in each packet in the IPv4 *Type of Service* (TOS) octet or the IPv6 Traffic Class octet (see Figure 4.33) is used to mark a packet to receive a particular forwarding treatment, or per-hop behavior, at each network node. A common understanding of the use and interpretation of this bit pattern is required for inter-domain use, multi-vendor interoperability, and consistent reasoning about expected aggregate behaviors in a network. Thus, the DiffServ Working Group of the *Internet Engineering Task Force* (IETF) has standardized a common layout for a six-bit field of both octets, called the DiffServ field (DS field) [90].

The first six bits of the DS field are used as a code point (*DiffServ Code Point* (DSCP)). These determine the *Per-Hop Behavior* (PHB) the packet sees at each node that usually consist of packet queuing and scheduling. PHBs define how traffic belonging to a particular behavior aggregate is treated at an individual network node. A two-bit field currently unused (CU) is reserved for future use. Depending on the first three bits of DSCP, eight precedence levels (classes) are available in DiffServ as shown in Table 4.8.

Table 4.8: Precedence levels of DiffServ based on bits 0, 1, 2 of DSCP

Bits 0, 1, 2 of DSCP	Precedence Level	Usage
111	7	link layer and routing protocol control
110	6	routing protocol control
101	5	Expedited Forwarding Class (EFC)
100	4	Assured Forwarding Class (AFC) 4
011	3	Assured Forwarding Class (AFC) 3
010	2	Assured Forwarding Class (AFC) 2
001	1	Assured Forwarding Class (AFC) 1
000	0	Best Effort Class (BEC)

Table 4.9: Proposal for mapping GPRS/UMTS classes to DiffServ classes

3GPP QoS Class	DiffServ Class	Reason
Conversational	EFC	low latency and jitter required
Streaming	AFC 4	low jitter required
Interactive	AFC 3	relatively low latency required
Background	AFC 2, AFC1 or BEC	only reliability required

4.4.6.2 Per-Hop Behaviors at Interior Routers

An interior router is any router that is not at the boundary of a DiffServ network domain. Since interior routers make up the majority of routers through which most IP packets pass, the complexity of the functions performed by interior routers must remain low. The DiffServ architecture recognizes this fact and mandates that only simple PHBs are implemented at interior routers.

4.4.6.3 Classification and Conditioning at Boundary Routers

The boundary router is located at the edge of a DiffServ network domain. This boundary router must perform sophisticated packet classification, metering, marking, policing, and shaping operations.

4.4.6.4 Bandwidth Broker

To make appropriate internal and external admission control decisions and to configure boundary devices correctly, each DiffServ domain is fitted with a bandwidth broker. The bandwidth broker performs admission control depending on network resources and configures boundary routers to take or drop a connection request. How the information on the traffic load situation at the network nodes is retrieved is not specified in the DiffServ specifications. The bandwidth broker might request information from the queues of all routers on one or several paths before admitting a request.

4.4.6.5 A Proposal for Mapping GPRS/UMTS Classes to DiffServ Classes

To integrate the QoS management functions in the radio network and the QoS functions in the IP core network, e.g. DiffServ, mapping rules have to be defined so that these functions can interwork efficiently. A proposal for mapping GPRS/UMTS classes onto DiffServ classes is presented in Table 4.9. In this proposal the Conversational class is mapped to the expedited forwarding class since it is a real-time application and requires both low delay and low delay jitter. The Streaming class is mapped to assured forwarding class 4 since it requires stringent delay jitter requirements. The Interactive class is mapped to the assured forwarding class 3 since it requires low latency but not as low as in the Conversational class. The Background class can be mapped to any lower DiffServ class.

5
EDGE and Evolution to GERAN

EDGE is a further development of the GSM data services HSCSD and GPRS and is suitable for circuit- and packet-switched services. The circuit-oriented part is the *Enhanced Circuit-Switched Data* (ECSD). The packet-oriented part is the *Enhanced General Packet Radio Service* (EGPRS). Applying modified modulation and coding schemes EDGE reaches very high raw bit rates of up to 69 kbit/s per GSM physical channel. If a user utilizes all 8 time slots in parallel, the theoretical maximum raw bit rate rises to 554 kbit/s. The maximum bearer bit rate achievable rises to about 384 kbit/s [59]. EDGE was introduced to the ETSI for the first time in 1997 for the evolution of GSM. After a successful feasibility study of the ETSI the standardization process for EDGE was initiated. Although EDGE was introduced for the evolution of GSM, this concept can be applied to increase the data rate in other systems. While the enhancements of the GSM for EDGE are mainly related to the physical and data link layer of the air interface, the evolution to *GSM/EDGE Radio Access Network* (GERAN) is characterized by modifications in the system and protocol architecture. Since EDGE-based radio access networks have the potential to complement UMTS networks, which will be based on the 3GPP standards for the *UMTS Terrestrial Radio Access Network* (UTRAN) and the 3G *Core Network* (CN), the same services should be accessible over UTRAN and GERAN over consistent interfaces with the same switching technology.

5.1 EDGE—Enhanced Data Rates for GSM Evolution

The integration of EDGE into GSM networks will be done gradually in order to minimize investment. So in the beginning EDGE transceivers will be able to process ordinary GSM signals and replace the older GSM transceivers. As result, GSM and EDGE signals will coexist in the same frequency band. Additionally it is desired to perform minimal changes in cell and frequency planning.

Since the network architecture of the GSM will remain similar for EDGE, the modifications at the air interface are depicted. The major changes in the GSM standard are made at the radio interface. To support higher data rates, a modulation scheme called *8-Phase-Shift-Keying* (8-PSK) is introduced which will not replace the current GMSK but coexist with it. With 8-PSK it is possible to provide a higher data rate, which is necessary to support bandwidth extensive data applications.

The modifications mostly concern the RLC/MAC layer and the physical layer. Since these protocols are placed in the MS and the BS, both have to be modified. In reality, the changes that have to be made comprise a new EDGE transceiver unit and software upgrades to the BSC, which then can handle standard GSM or GPRS traffic and will automatically switch to EDGE mode when needed.

The core of EDGE is the *Link Quality Control* (LQC) mechanism that allows the adaptation of the *Modulation and Coding Schemes* (MCSs) to a changing radio

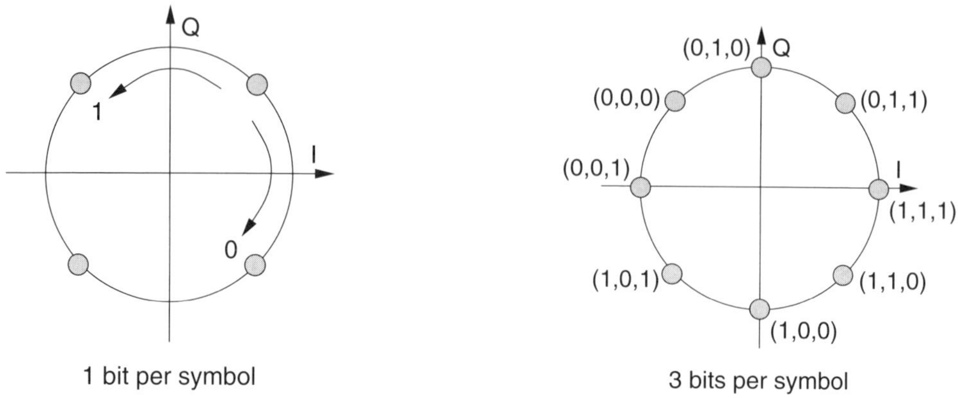

(a) GMSK modulation (b) 8-PSK modulation

Figure 5.1: I/Q-diagram of GMSK and 8-PSK modulation

link quality. Although *Link Adaptation* (LA) is already possible within the GPRS standard, higher CSs are not supported by the actual equipment and will probably only be introduced together with EDGE functionality.

Additionally, a type-II-hybrid ARQ (soft ARQ) scheme is introduced. Soft information is stored during retransmissions to enable *Incremental Redundancy* (IR). The RLC/MAC protocol structure and retransmission mechanism proposed for EGPRS are based on the GPRS standard. An LLC frame is divided into a number of RLC blocks. The average duration of an RLC block is 20 ms, whereas its information content ranges from 176 to 1184 bit, depending on the MCS used. For error detection, the RLC blocks contain a CRC field. Upon reception of an RLC block, the receiver checks the CRC and determines if retransmission of the RLC block is necessary. To give a more detailed introduction, in the following the modulation scheme 8-PSK used for EDGE is described and compared with GMSK currently used for GSM. Additionally the new MCSs are presented. Then the main characteristics of the LQC mechanisms are given in Section 5.1.3, comprising LA and IR.

Finally the RLC/MAC protocol is adapted to the higher data rates that have to be supported. The RLC parameters *Sequence Number Space* (SNS), *Window Size* (WS) and consequentially the *Block Sequence Number* (BSN) field and the *Received Block Bitmap* (RBB) structure are adapted (see Section 3.5.3).

5.1.1 8-PSK Modulation versus GMSK Modulation

The modulation scheme that is used in GSM is called *Gaussian Minimum Shift Keying* (GMSK). In GPRS the same modulation type will be used and high bit rates are achieved through multislot operation. To allow higher bit rates within EGPRS, the *8-Phase-Shift-Keying* (8-PSK) modulation scheme is introduced in addition to GMSK [59, 102]. The differences between GMSK and 8-PSK are illustrated in an I/Q diagram as shown in Figure 5.1.

Within GMSK transmitting a 0-bit or a 1-bit is represented by incrementing the phase with $\pm 1/2\,\pi$. Every symbol that is transmitted represents one bit, which means that symbol rate and bit rate are equal. The symbol rate of 271 kbit/s is divided over

Table 5.1: EGPRS modulation and coding schemes and coding parameters

MCS	CR	HCR	Data per radio block [bit]	Fam.	RLC info [byte]	Data rate [kbit/s]
9 (8-PSK)	1.0	0.36	2×592	A	2×74	59.2
8 (8-PSK)	0.92	0.36	2×544	A	2×68	54.4
7 (8-PSK)	0.76	0.36	2×448	B	2×56	44.8
6 (8-PSK)	0.49	1/3	592 *(544+48)*	A	74	29.6 (27.2)
5 (8-PSK)	0.37	1/3	448	B	56	22.4
4 (GMSK)	1.0	0.53	352	C	44	17.6
3 (GMSK)	0.80	0.53	296 *(272+24)*	A	37	14.8 (13.6)
2 (GMSK)	0.66	0.53	224	B	28	11.2
1 (GMSK)	0.53	0.53	176	C	22	8.8

eight time slots. After burst formatting (see Section 2.1.7.1) a data rate of 22.8 kbit/s per time slot results. Via different channel coding schemes this results in different net bit rates.

Within 8-PSK three consecutive bits are mapped onto one symbol in the I/Q-plane. With the same symbol rate as in GMSK of 271 kbit/s, bit rates of up to 813 kbit/s can be achieved. With burst formatting, this results in a bit rate of 69.2 kbit/s per time slot. 8-PSK modulation still has the GSM spectrum mask and leaves the burst duration unchanged, which offers the possibility of using both modulation schemes next to each other. However, 8-PSK modulated data is less robust if the channel conditions are bad. The advantage of the new modulation scheme is support for higher data rates under good channel conditions, and to reuse at the same time the channel structure of the GPRS system. Since the *Bit Error Ratio* (BER) depends significantly on the channel conditions, *Link Quality Control* (LQC) is of major importance for a high throughput in EGPRS.

LA as part of LQC gives the possibility of changing to another modulation scheme or channel coding, when the channel quality changes. Through the adaptation to the radio channel conditions it is possible to offer the highest data rates in good propagation conditions close to the site of the BSs, whereas under lower quality conditions a more robust coding scheme is selected.

5.1.2 Modulation and Coding Schemes

The additional use of 8-PSK modulation enables the introduction of new *Modulation and Coding Schemes* (MCSs). The possible MCSs are depicted in Table 5.1. MCS-1 to MCS-4 use GMSK as the modulation scheme and nearly equal the coding schemes used in GPRS, providing data rates up to 17.6 kbit/s. MCS-5 to MCS-9 use the previously described 8-PSK modulation scheme enhancing the data rate up to 59.2 kbit/s per time slot. The MCSs are classified in three different MCS families consisting of different basic payload sizes. Figure 5.2 gives an overview of how an RLC block is created depending on the MCS family. For instance, MCS-9 comprises four basic units of 37 octets each, whereas MCS-6 combines two basic units and MCS-3 only one. Through transmitting a different number of payload units within a 20 ms radio block duration, different code rates are achieved, resulting in bit rates of 8.8 kbit/s up to 59.2 kbit/s per time slot. According to the link quality, an initial MCS is selected

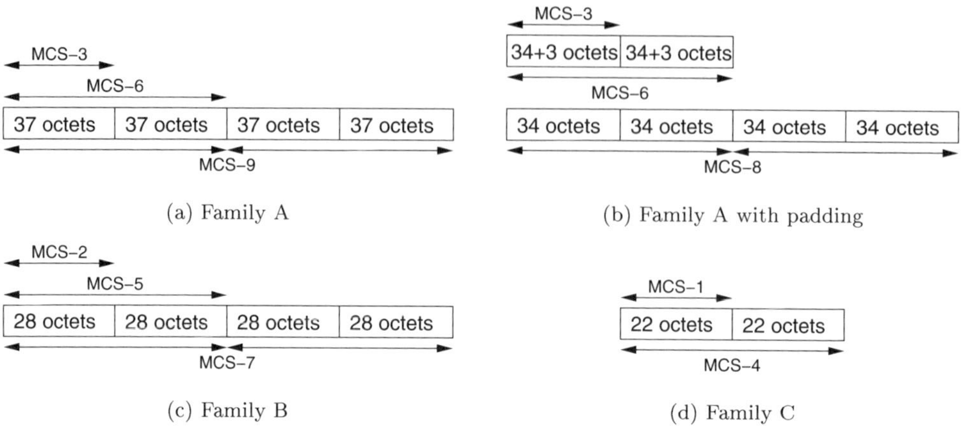

(a) Family A (b) Family A with padding

(c) Family B (d) Family C

Figure 5.2: Basic block sizes and MCS families

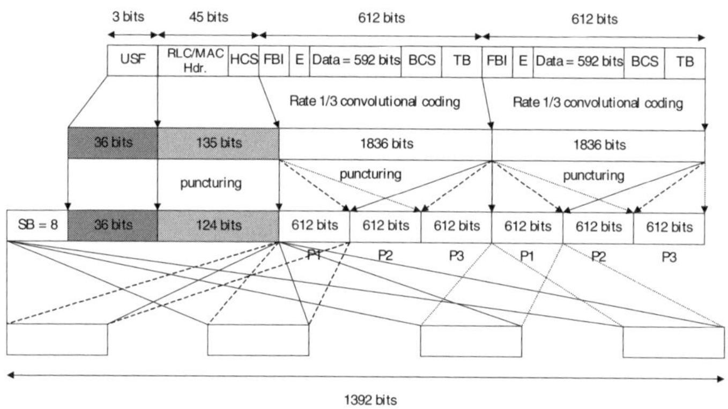

Figure 5.3: Coding and puncturing for MCS-9; uncoded 8-PSK, two RLC blocks

for each RLC block. For retransmissions, it is only allowed to use the same MCS or to switch to another MCS of the same MCS family.

The following figures show the coding and puncturing used in Family A (MCS-9: Figure 5.3, MCS-6: Figure 5.4 and MCS-3: Figure 5.5). They differ in the amount of redundancy used for error correction, which determines accordingly the payload size.

Note that using MCS-7, MCS-8 and MCS-9, two RLC PDUs are transmitted within 20 ms. Both PDUs are interleaved over only two bursts, whereas their common header is interleaved over all four. Using a lower MCS, the RLC PDU (data and header) is interleaved over all four bursts. Figure 5.3 shows a MCS-9 RLC/MAC block consisting of two RLC PDUs of 592 bit (74 octets) payload and separate header containing the FBI field, extension field and the BSN (not pictured) to identify the RLC PDU within a TBF. The two RLC PDUs do not have to be in sequence; for example the first PDU can be a new RLC PDU and the second a retransmission of a previously sent RLC PDU. If there are no further RLC blocks to send, no retransmissions requested and no pending retransmissions foreseen, the second payload will be padded.

Initially, the RLC data block will be encoded with Puncturing Scheme PS-1, whereas PS-2 and PS-3 are used for retransmissions, then enabling the use of IR

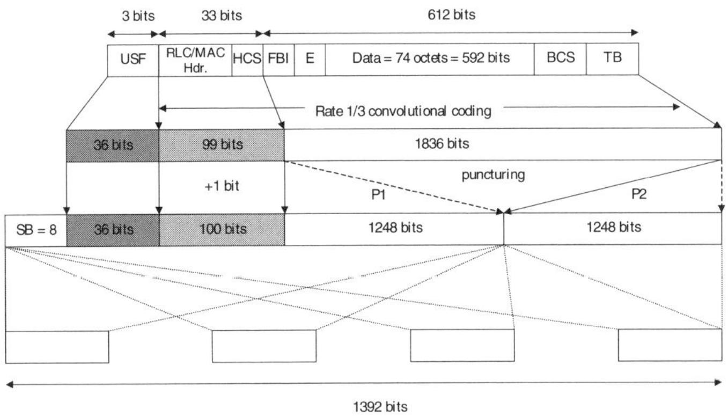

Figure 5.4: Coding and puncturing for MCS-6; rate 0.49 8-PSK, one RLC block

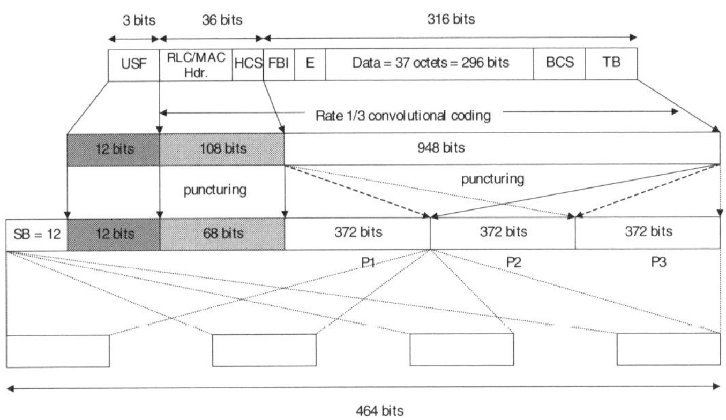

Figure 5.5: Coding and puncturing for MCS-3; rate 0.80 GMSK, one RLC block

by using the different punctured information in PS-1, PS-2 and PS-3. The EDGE standard foresees an LA algorithm for the dynamic selection of MCSs to adapt to the quality of the radio link. This includes measurement and reporting the quality of the downlink and demands the indication of new modulation schemes (see Figure 5.6).

5.1.3 Link Quality Control

Link Quality Control (LQC) is a term used for techniques to adapt the channel coding of the radio link to the varying channel quality. Different modulation and coding schemes are optimal during different situations, depending on the link quality. The LQC used for EDGE is performed through the techniques of

- *Link Adaptation* (LA)
- *Incremental Redundancy* (IR)

LA provides a dynamic switching between coding and modulation schemes, so that the highest throughput (e.g., maximum user bit rate) according to the time-varying link quality (e.g. C/I) is achieved. With IR, information is first sent with very little coding. If decoding is successful, this will yield a very high user bit rate or

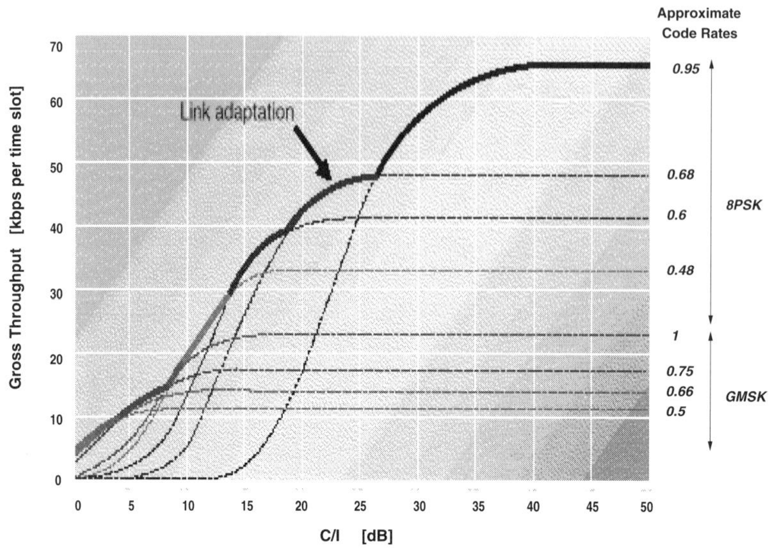

Figure 5.6: Link adaptation algorithm [60]

throughput. However, if decoding is unsuccessful, then additional coded bits are sent until decoding is successful. The more coding bits that have to be sent, the less the resulting bit rate and the higher the delay. As a result of the link quality control, a low radio link quality will not cause a dropped data transfer, but only give a reduced bit rate for the user. Furthermore, a tight frequency plan (with a small cluster size, say 3) can be introduced while still providing high data rates for packet data services.

5.1.3.1 Link Adaptation (LA)

An LA scheme is proposed to maximize the user bit rate on the link through estimation of the time varying link quality in EGPRS and respectively adapt the most appropriate modulation and coding scheme. Through measurements of the link quality, the predictive algorithm estimates the performance of the currently used scheme and decides whether another MCS would perform better than the currently used one. The algorithm chooses the MCS with the best performance for the currently measured radio link quality and therefore for the expected link quality during the next bursts (it is assumed that the link quality can vary with time and location). During active TBFs the most appropriate MCS is chosen on the basis of the channel quality measurements concerning the used MCS family. For compatibility reasons it is not allowed to switch the chosen MCS family during retransmissions, otherwise too much padding would decrease efficiency.

In order to perform the most efficient LA, measurements of the quality of the radio channel are the basis of the choice of the MCS. The operation of link quality control relies on measurements of the link quality, depending also on C/I values, frequency errors, time dispersion, interleaving gain due to frequency hopping, velocities and bit error probability.

5.1.3.2 Incremental Redundancy

Operating in RLC acknowledged mode, retransmissions are possible, when requested, or when capacity is available for pending retransmissions (RLC blocks not yet acknowledged). An ARQ scheme ensures an error-free data transmission by retransmitting erroneous blocks. When an error is detected in a block, all information about the block is usually discarded. The basic idea of IR is soft combining, i.e., not to forget what was already received in the erroneous transmission. Instead, the soft bit information from the erroneous blocks is saved and combined with the information received with the next (erroneous) retransmission. That way the *Block Error Ratio* (BLER) is reduced.

With IR mode PS-1 is used at first, and if decoding is unsuccessful, PS-2 will be used, etc. The IR functionality saves soft information of PS-1 and performs a joint decoding with PS-2 after the received retransmission with PS-2. A *Synchronization downlink Burst* (SB) is used to indicate which code rate and which puncturing scheme is used. Information about link quality control is given in [16], [17], and [19]. Through measurements of the link quality every 60 ms (example inter-arrival time of the measurement report), the most suitable MCS family is chosen, based on the mean and variance of the C/I. For the following explanation, MCS family A is assumed. All assumptions apply also for the other families. For retransmissions three possible scenarios are drawn. The first one assumes that the link quality remains stable during the retransmissions. Then the RLC data blocks can be retransmitted with the originally used MCS. If the link quality worsens, an MCS downgrading becomes necessary, otherwise if the link quality improves, an MCS upgrading is possible. In the following implementation, proposals for upgrading and downgrading actions are described.

Upgrading When performing an MCS upgrade, retransmissions can be executed using the old MCS, allowing the use of IR by using different puncturing schemes (PS-1, PS-2, PS-3). That way, the BLER is reduced by using information from the different puncturing schemes. Otherwise no IR is possible.

Downgrading If the channel quality worsens, an MCS downgrading is necessary to provide a low BLER through higher protection of the bits. Thus, the retransmissions are carried out using a lower MCS, rejecting the possibility of the use of IR because of the incompatible puncturing of the different MCSs. For instance, when MCS-6 is chosen, when previous RLC blocks were sent in MCS-9, no coding problem occurs since MCS-9 comprises two individual RLC blocks consisting of two basic blocks each. MCS-6 also comprises two basic blocks, therefore only the coding, puncturing and interleaving will change and not the number of RLC blocks distinguished by separate BSNs. Retransmission of a previous MCS-9 RLC block results in a retransmission of two separate RLC blocks in MCS-6.

Assuming MCS-6 and downgrading to MCS-3, the payload data has to be deblocked into two subblocks (also called split blocks) and the header duplicated, creating two MCS-3 RLC blocks with the same BSN. Both MCS-3 blocks are retransmitted consecutively with the split bit set to 1 for the first and 2 for the second MCS-3 block to allow accurate assembly at the receiver side requiring the error-free arrival of both the first and second subblocks. Performing downgrading, all consecutive RLC blocks will be sent using the lower MCS, remaining in this MCS even if changed channel quality allows upgrading. This applies only for retransmissions, whereas the newly segmented LLC frame will be sent with the actual MCS, taking advantage of the better channel quality and the transmission capacity of the higher MCS.

Note that it is not possible to acknowledge the subblocks separately, hence the receiver has to wait until both subblocks have arrived before they enter the receiver window. The RLC block, consisting of two subblocks, is acknowledged positively if both subblocks have arrived error-free, otherwise both subblocks have to be requested by using the BSN of the former MCS-6 PDU.

5.1.4 Flow Control Modifications for EDGE

To support higher data rates in an efficient way the flow control function in the RLC/MAC protocol has to be modified. The *Window Size* (WS) is increased from 64 in GPRS to the range from 64 to 1024 in EGPRS. Consequentially the *Sequence Number Space* (SNS) is adapted from 128 to 2048 and the *Block Sequence Number* (BSN) is increased from 7 bit to 11 bit. To be able to maintain the selective achnowledgement function for this large SNS the *Received Block Bitmap* (RBB) is segmented into a *First Partial Bitmap* (FPB) and *Next Partial Bitmap* (NPB) controlled by the *Extended Polling Function* in EGPRS.

In EGPRS downlink TBFs, an additional poll bit is added to the S/P field in every downlink RLC block so that the network can request either an FPB segment or an NPB segment. While for an FPB the bitmap starts with SSN = V(Q) + 1, for the NPB an additional offset is specified by the *Partial Bitmap Sequence Number* (PBSN) that is stored at the receiver and that can be determined by the sender after reception of the NPB.

Additionally compressed bitmaps are introduced for so-called *Reported Bitmaps* (RBs). If the window size is larger than the number of bits available for the bitmap, one-dimensional runlength coding is carried out.

5.1.5 The EDGE Compact Concept

EDGE Compact [55] is a variant of EGPRS that is designed for deployment in narrow frequency bands as in the US for EDGE-136 systems. Here the requirement is deployment in less than 1 MHz of spectrum. In GSM networks continuous transmission with constant transmit power in the BCCH frequency is assumed, so that the BCCH needs a relatively large reuse factor to ensure reliable reception of the system information on the broadcast and common control channels. Therefore the spectral efficiency of the GSM network starts to decrease if the allocated frequency band is small.

To make GSM/EDGE networks deployable in only two times 600 kHz of spectrum by using three carriers in a 1/3 reuse pattern for data traffic, a new control channel structure is introduced, which is based on the discontinuous time rotating BCCH, enabling significantly lower reuse factors for the BCCH. Inter base station time slot synchronization is used, where in time slots of common control channels no signaling data is transmitted within the same time group (see Figure 5.7).

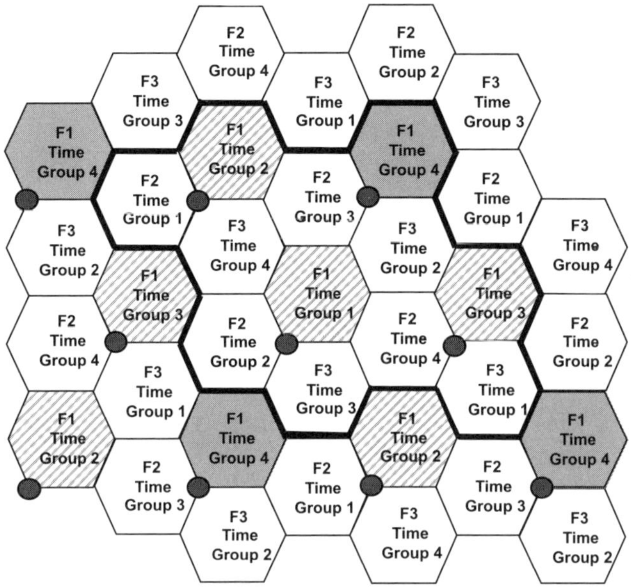

Figure 5.7: Three-carrier EDGE Compact concept

5.2 GERAN—GSM/EDGE Radio Access Network

In the context of the standardization activities of the *3rd Generation Partnership Project* (3GPP) for the *Universal Mobile Telecommunication System* (UMTS) new concepts in the system and protocol architecture for the access network, called the *UMTS Terrestrial Radio Access Network* (UTRAN), have been developed. While in GPRS/EDGE networks a logical link is defined between MS and SGSN (see Section 3.4), so-called *Radio Access Bearers* (RABs), which are managed by the *Radio Resource Control* (RRC), are defined between the MS and the UTRAN. Hence, the locations of several functions have been shifted from the SGSN that is already part of the GPRS core network to the *Radio Network Controller* (RNC) that is part of the UTRAN. A *3rd Generation Serving GPRS Support Node* (3G-SGSN) no longer comprises logical link management functions. Additionally header and data compression, ciphering and radio resource management are performed by *Radio Network Controller* (RNC) protocols in *UMTS Terrestrial Radio Access Networks* (UTRANs).

The interface between the *Radio Access Network* (RAN) and the core network is called I_u and comprises two component interfaces, the I_{u-cs} for circuit-switched services and the I_{u-ps} for packet-switched services. To be able to operate *GSM/EDGE Radio Access Networks* (GERANs) in parallel with UTRANs and to provide a similar set of services through both radio access technologies, the system and protocol architecture of GPRS/EDGE was adapted to the *UMTS Terrestrial Radio Access Network* (UTRAN) so that a GERAN can be connected to the I_u interface and use the same core network.

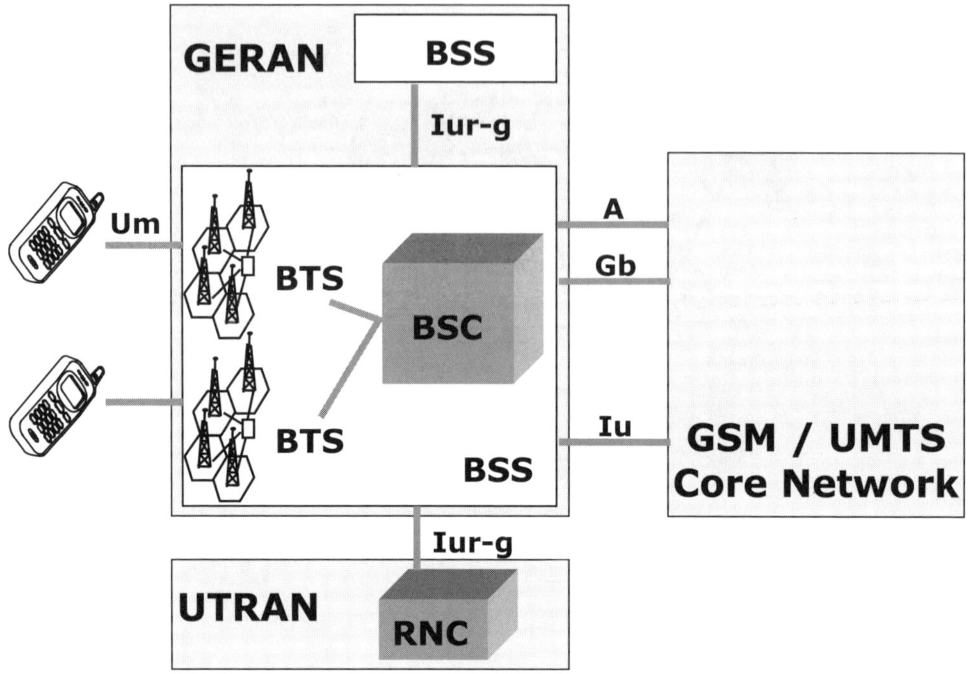

Figure 5.8: GERAN architecture

5.2.1 GERAN System Architecture

After the introduction of the G_b interface for GPRS, all following standard releases were based on two different core network interfaces, the interface between BSS and MSC (A) for circuit-switched services and the interface between BSS and SGSN (G_b) for packet-switched services. For GERAN [18] the UMTS I_u was adapted and the GERAN architecture is aligned with the UTRAN architecture. Figure 5.8 shows the GERAN architecture with the interfaces A and G_b that coexist with the I_u interface to realize backward compatibility. This architecture conforms with the general UMTS architecture [26], where the functional split between the *Access Stratum* and the *Non-Access Stratum* is defined (see Figure 5.9). The Access Stratum only provides services related to the transmission of data over the radio interface and the management of the radio interface. The Access Stratum offers services to the Non-Access Stratum through the following *Service Access Points* (SAPs):

General Control (GC) SAPs are used to enable the core network to provide information and to give commands that are not related to specific users or sessions.

Notification (Nt) SAPs are used to broadcast data, e.g., paging messages to identified users.

Dedicated Control (DC) SAPs are used to establish and release connections with specific user equipment and to exchange information related to these connections.

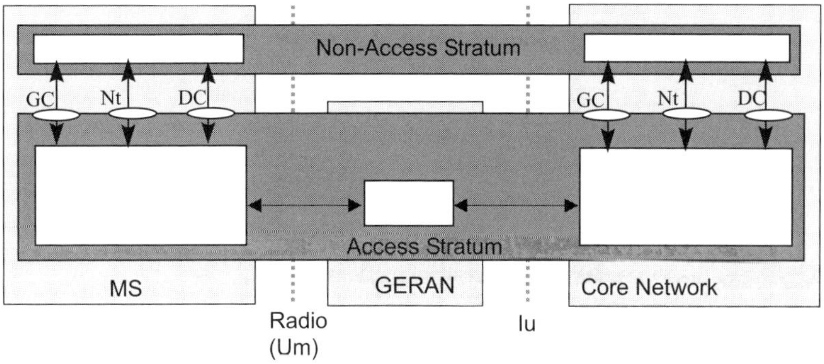

Figure 5.9: Access Stratum and Non-Access Stratum applied to GERAN

5.2.2 GERAN Protocol Architecture

In GPRS networks the protocols SNDCP and LLC in the SGSN are providing functions that are related to the *Radio Access Network* (RAN) for the user plane and the control plane. Because of the functional split between core network and RAN in GERAN, these functions are shifted to the *Base Station Controller* (BSC). While in GPRS networks LLC offers several *Service Access Points* (SAPs) for signaling and user data of different service classes (see Section 3.4), in GERAN the *Radio Resource Control* (RRC) is defined that manages different *Radio Access Bearers* (RABs) and configures the protocol stacks of the user plane (see Figure 5.10) and control plane in the GERAN. Four types of RABs are defined with different QoS requirements (see also Section 4.4.5):

- Conversational

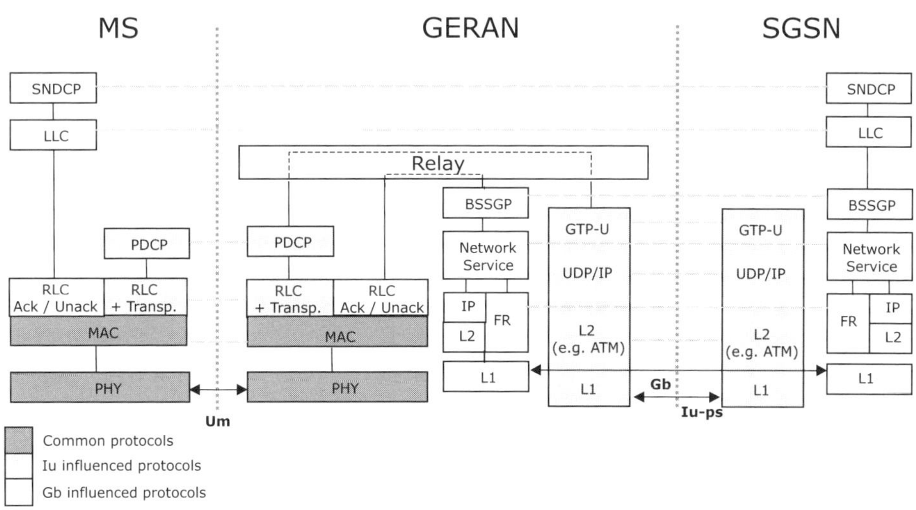

Figure 5.10: The GERAN user plane

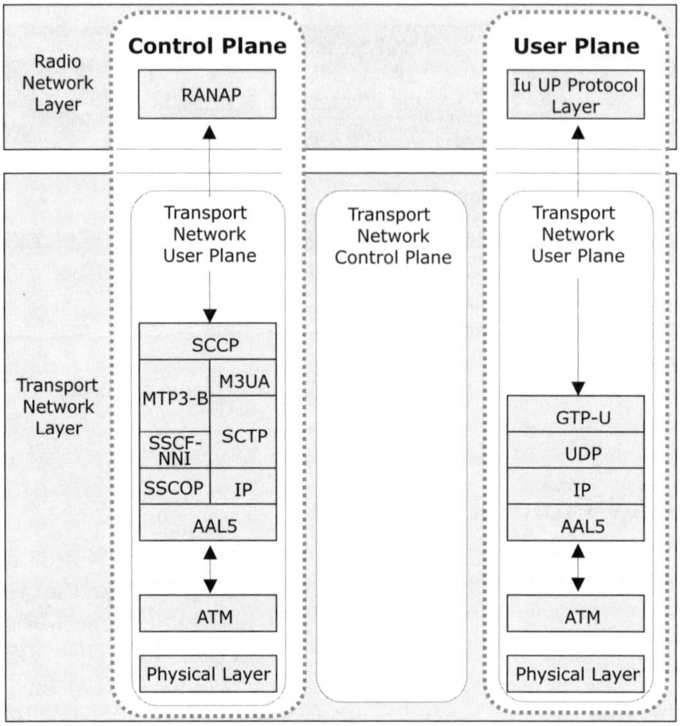

Figure 5.11: The GERAN control plane at the I_u interface

- Streaming
- Interactive
- Background.

Conforming with UMTS, between BSC and SGSN the *GPRS Tunneling Protocol (User Plane)* (GTP-U) (see Section 3.1) is used to tunnel user data (see Figure 5.10) and extended Signaling System Number 7 protocols are used to carry signaling messages [18]. This signaling transport is used by the I_u-signaling protocol *Radio Access Network Application Part* (RANAP) that is responsible for the management of RABs at the I_u and for the management of the I_u interface itself. The protocols of user and control plane at the I_u are shown in Figure 5.11. As the network service *Asynchronous Transfer Mode* (ATM) technology can be applied. However, more transport options will be possible in GERAN.

In the following the modifications of the functions and protocols of the radio interface that have been aligned with the *UMTS Terrestrial Radio Access Network* (UTRAN) protocols are introduced.

5.2.2.1 Radio Resource Control (RRC)

The *Radio Resource Control* (RRC) is part of the control plane and is based on the UTRAN RRC protocol [21]. It performs signaling functions between MS and

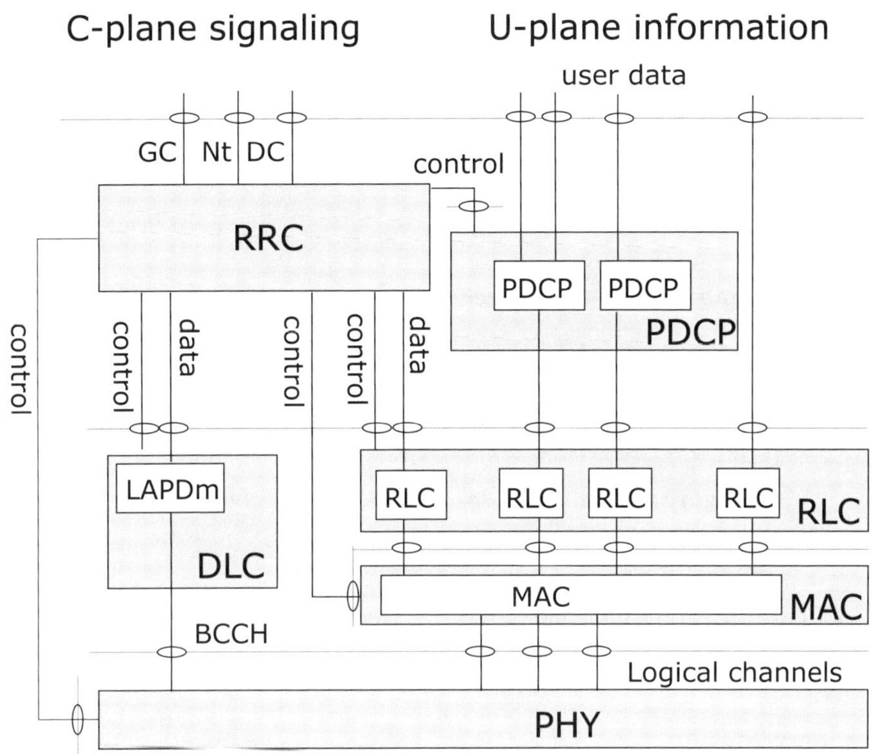

Figure 5.12: Protocol configuration performed by the Radio Resource Control

GERAN. Besides the broadcast of system information related to the Access Stratum and the Non-Access Stratum, the RRC is responsible for the management of RRC connections, which are logical contexts between MSs and GERAN. Additionally it establishes, reconfigures and releases RABs that are used for user data transmission and signaling. In this configuration function the protocol options of the user plane are defined in order to meet the QoS requirements of different service classes (see Figure 5.12). While a user plane bearer is set up, when an RAB is requested by the core network, control plane bearers are established, when an RRC connection is set up by an MS.

5.2.2.2 Packet Data Convergence Protocol (PDCP)

The PDCP comprises the functions performed by the SNDCP in GPRS networks (see Section 3.3). A transparent mode and a non-transparent mode are defined. In non-transparent mode header adaptation can be performed. Two different header adaptation algorithms are defined:

Header Compression: Transport and network layer headers are compressed based on the *Robust Header Compression* (ROHC) specification [37] (see Section 6.6.3.3).

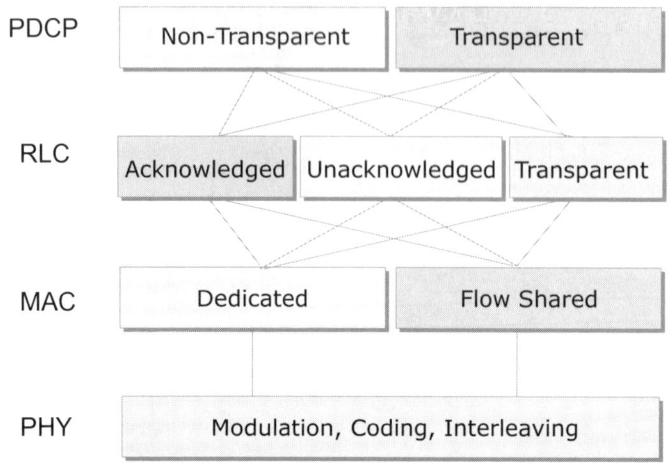

Figure 5.13: PDCP, RLC and MAC modes of operation

Header Removal: Transport and network layer headers are completely removed. Based on information submitted at call setup and based on information from lower layers the receiving entity can regenerate the headers.

5.2.2.3 Radio Link Control

The RLC protocol in GERAN is extended by a transparent mode for Conversational bearers (see Figure 5.13). In this mode RLC has no functionality. Incoming SDUs are directly forwarded to the MAC layer. For Streaming, Interactive and Background bearers the acknowledged and unacknowledged mode of RLC can be used (see Section 3.5.3). Since the LLC protocol is removed from the SGSN, ciphering in GERAN is performed by RLC in non-transparent mode and by MAC in transparent mode.

5.2.2.4 Medium Access Control

The MAC protocol for packet data is based on the GPRS MAC protocol introduced in Section 3.5.4. A dedicated mode is added to map delay-critical data flows on dedicated channels. Since dedicated channels were not managed by the (E)GPRS RLC/MAC protocol, the function of mapping logical channels on physical channels has been modified (see Figure 5.14). While the management of dedicated channels is performed by RRC, shared channels are managed by MAC. For shared channels the CSs/MCSs of (E)GPRS are used. If dedicated channels are used for a RAB, the CSs/MCSs of (E)CSD are used. Additionally a ciphering function is provided, if RLC is operating in transparent mode.

5.2.2.5 Basic Physical Subchannels

In convergence with UMTS, where channels are either *dedicated channels* or *shared channels*, two types of *Basic Physical Subchannels* (BPSCHs) are defined in the

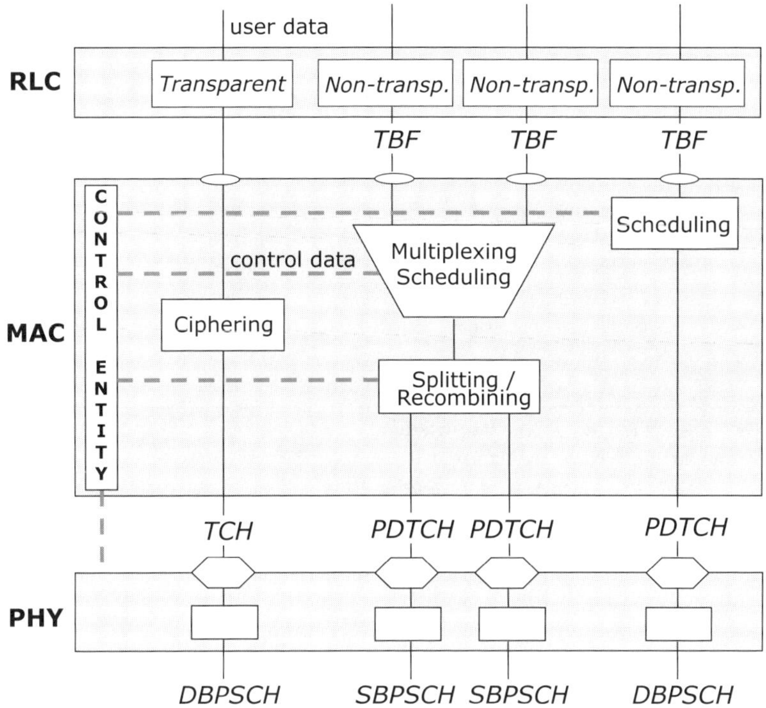

Figure 5.14: MAC over dedicated and shared channels

physical layer of GERAN additionally to the GSM physical channels known from GSM/EDGE (see Section 2.1.7.2).

The *Dedicated Basic Physical Subchannel* (DBPSCH) is assigned to only one user, while for packet-switched services the *Shared Basic Physical Subchannel* (SBPSCH) is defined. As the PDCH in (E)GPRS, the SBPSCH uses a 52-frame multiframe structure and offers 12 radio block periods per 52-frame multiframe that can be used by the GPRS logical channels (see Section 2.2.5.1).

5.2.2.6 Radio Access Bearer Realization

As introduced in Section 5.2.2.1 the RRC protocol configures the protocol options of PDCP, RLC, MAC and PL to meet the QoS requirements of different RABs. These options comprise the modes of PDCP, RLC and MAC, the type of the BPSCH and the traffic channel parameters, namely interleaving, modulation and coding schemes. Table 5.2 shows the configuration of some protocol options for the four RAB types.

5.2.3 Core Network Evolution

The GPRS/UMTS core network specified in 3GPP Release 99 specifications is split into two subnetworks, the *CS domain* for circuit-switched services and the *PS domain* for packet-switched services (see Figure 5.15). The CS domain is based on *Time Division Multiplex* (TDM) switching technology known from *Public Switched Telephony*

Table 5.2: Radio Access Bearer realization

RAB	PDCP	RLC	MAC	BPSCH type
Conversational	NT (HR)	T	Dedicated	DBPSCH/F or /H
	NT (HC or NA) or T	Uack	Dedicated	DBPSCH/F or /H
Streaming	NT (HC or NA) or T	Uack	Dedicated	DBPSCH/F or /H
	NT (HC or NA) or T	Ack	Dedicated	DBPSCH/F or /H
	NT (HC or NA) or T	Uack	Shared	SBPSCH/F or /H
	NT (HC or NA) or T	Ack	Shared	SBPSCH/F or /H
Interactive	NT (HC or NA) or T	Ack	Dedicated	DBPSCH/H or /H
	NT (HC or NA) or T	Ack	Shared	SBPSCH/H or /H
Background	NT (HC or NA) or T	Ack	Dedicated	DBPSCH/H or /H
	NT (HC or NA) or T	Ack	Shared	SBPSCH/H or /H

Ack/Uack — Acknowledged / Unacknowledged
T/NT — Transparent / Non-Transparent
HC/HR/NA — Header Compression / Header Removal / No Adaptation

Networks (PSTNs) and Signaling System Number 7 protocols for signaling. The PS domain is an evolved GPRS core network using IP transport technology. Signaling interfaces of the PS domain between the *GPRS Support Nodes* (GSNs) and *Mobile-services Switching Center* (MSC) and *Home Location Register* (HLR) are based on Signaling System Number 7 protocols. The 3GPP vision to create standards for so-called *All-IP networks* at the end of the 1990s was driven by three trends in industry and academia. Because of the majority of the data traffic worldwide being IP traffic, motivation has increased to adapt non-IP traffic to IP to reduce cost with a common transport. Additionally more and more applications rely on IP, so that IP can be seen as a service enabler for IP multimedia services supporting both telephony and data applications. The third trend can be seen from a global perspective, especially regard-

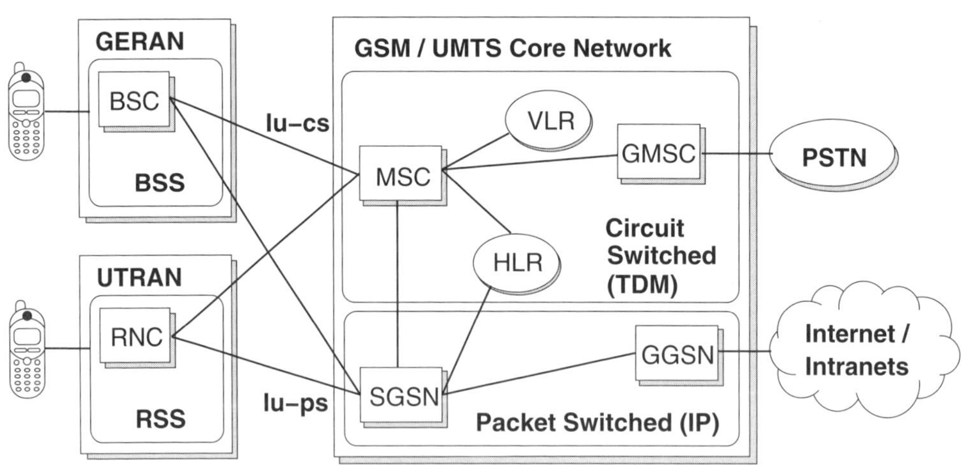

Figure 5.15: Core Network of 3GPP Release 99

ing the 3GPP2 standardization that has set the direction of new IP-based standards for cdma2000 voice, data and multimedia.

The evolution towards an All-IP network, also called IP-Multimedia network, within the 3GPP is occurring in three steps corresponding to Releases 4, 5 and 6. In the following the main concepts of the core network evolution in the framework of these 3GPP releases is explained briefly. While the standardization of Releases 4 and 5 is ongoing, no specifications for Release 6 exist yet. Therefore the concepts presented for 3GPP Release 6 are only reflecting the visions of parts of the 3GPP [22].

5.2.3.1 3GPP Release 4

In 3GPP Release 4 the *Bearer Independent Circuit Network* (BICN) [5] is introduced as an optional new architecture that is alternative to the conventional TDM switching technology in the CS domain (see Figure 5.16). Basic new concepts are modified neither in the PS domain nor at the radio interface and service set. The aim of this option is to enable operators to use IP or ATM as a common transport to reduce cost. Using *Media Gateways* (MGWs) at the edges of the CS domain, circuit-switched voice traffic from the *Radio Access Network* (RAN) or the PSTN is adapted to *Voice over IP* (VoIP).

The nodes *Mobile-services Switching Center* (MSC) server, *Gateway Mobile-services Switching Center* (GMSC) server and MGW are introduced. The GMSC server and MSC server provide the call control and mobility management functions of a GSM/UMTS (G)MSC, and the MGW provides the bearer control and transmission resource functions. The media gateway contains the stream manipulating functions for the adaptation of circuit-switched voice traffic to VoIP. The GMSC server and MSC servers are connected to the MGW using H.248 [75], which has been jointly developed within the ITU-T and the IETF, and supports a separation of call control entities from bearer control entities, and a separation of bearer control entities from

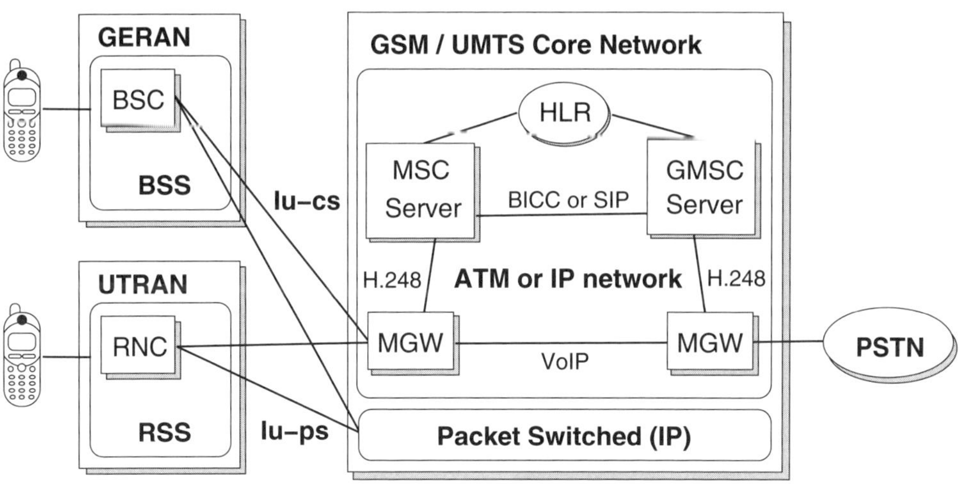

Figure 5.16: Core network of 3GPP Release 4

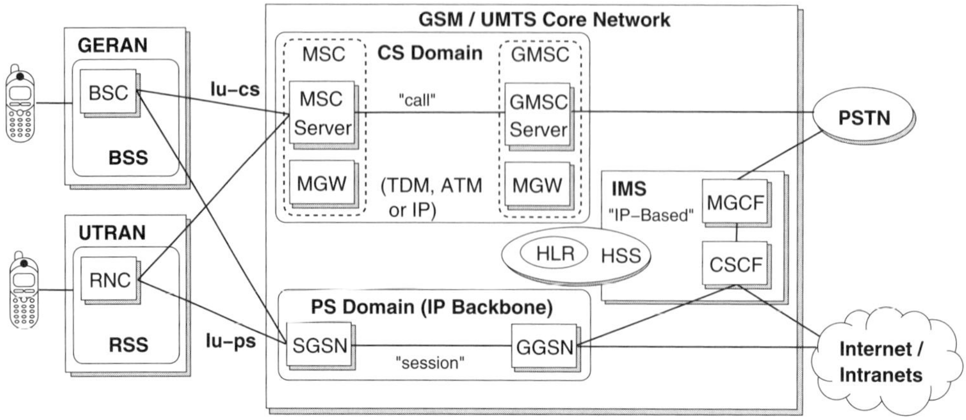

Figure 5.17: Core network of 3GPP Release 5

transport entities. The MSC servers and GMSC servers are connected using *Bearer Independent Call Control* (BICC) [5] or *Session Initiation Protocol* (SIP) (see Section 6.6.1) protocols. There may be a number of call control transit nodes between the MSC server and GMSC server. The MGWs are connected using RTP/UDP/IP (see Section 6.6).

To make the transport of GSM/UMTS speech over IP not noticeable for the end-user, the IP network needs careful engineering to meet the performance requirements and probably requires *Multi Protocol Label Switching* (MPLS) [99] or other techniques to manage IP QoS. MPLS prefixes the carried PDUs with a label to provide deterministic paths.

5.2.3.2 3GPP Release 5

While the core network modifications in Release 4 are related to IP transport, Release 5 aims to offer operators the possibility of increasing revenue by the use of IP-oriented services. The aim is to provoke a user migration from conventional users of CS speech served by the CS domain to users accessing enhanced IP-based services using the PS domain. This is realized by the *IP Multimedia Subsystem* (IMS) that overlays the existing architecture (see Figure 5.17). In order to achieve access independence and to maintain a smooth interoperation with wireline terminals across the Internet, the IP multimedia subsystem attempts to be conformant to IETF "Internet standards". Therefore, the interfaces specified conform as far as possible to IETF "Internet standards" for the cases where an IETF protocol has been selected, e.g., SIP.

The IP multimedia subsystem utilizes the PS domain to transport multimedia signaling and bearer traffic. For call control the functions *Call State Control Function* (CSCF) and *Media Gateway Control Function* (MGCF) are introduced. The CSCF acts as an entry point for the signaling of incoming calls. It performs the necessary database queries for location management and routes the call to its destination. To perform its tasks, the CSCF interacts with a new integrated database called *Home Subscriber Server* (HSS), which provides user profile information, either directly or

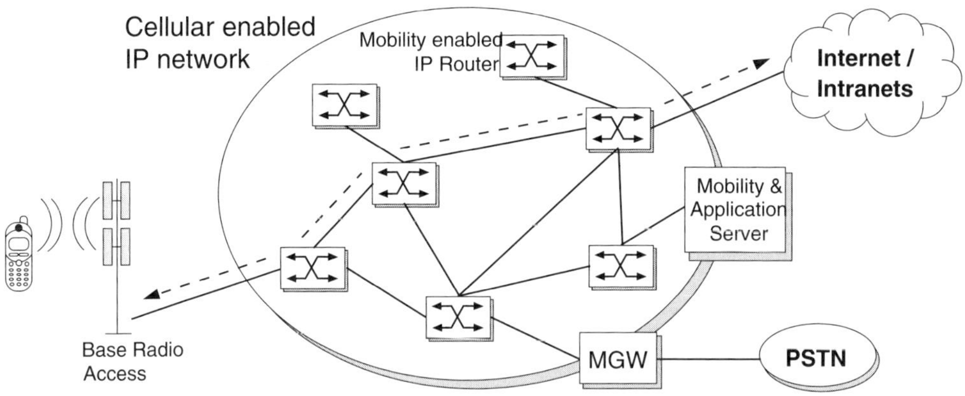

Figure 5.18: Candidate core network of 3GPP Release 6

via servers. The HSS is a superset of the HLR functionality (see Section 2.1.3), but differs in that it needs to also communicate via new IP based interfaces. The MGCF is responsible for the interworking with the PSTN/CS domain.

The introduction of the IMS in UMTS and evolved GSM networks will entail three problems to be solved. With the migration from CS voice users to SIP users, VoIP is applied over the air interface and the delay and protocol overhead for voice frames has to be reduced (see Section 6.6.3). The use of IP-based services will increase the need for an expanded IP address space. The introduction of IPv6 in the IMS can solve this problem. Finally roaming functions for home CSCF platforms and external CSCF platforms have to be introduced.

The optional deployment of 3GPP Release 5 appeals only to those operators wishing to add value to UMTS through new services and to operators with a strategic direction towards IP services.

5.2.3.3 Beyond Release 5

The logical conclusion of All-IP is to fully merge the cellular network with the IP network. Today cellular networks rely on specific protocols and nodes to manage mobility, security and other cellular requirements. From the perspective of the IETF and the 3GPP2 these functions could become integral and intrinsic to the IP network. The aim is to have one simple network that can support all access types. This radical simplification would lead to a cellular enabled IP network (see Figure 5.18) that could be composed of mobility enabled IP routers supported by mobility and application servers and media gateways. Wireless specific elements could be limited to a thin layer at the base station. Candidate technologies are the IETF specifications Mobile IPv6 [79] and *Authentication, Authorization and Accounting* (AAA) [87].

6
Evolving Applications and Traffic Models for GPRS and 3G

In parallel with GSM evolution, the data applications performed by mobile users will evolve. In the first phase of the GSM evolution, where the data services *Circuit Switched Data* (CSD) and GPRS are available, applications based on the *Wireless Application Protocol* (WAP) [117] running on smart phones and *Personal Digital Assistants* (PDAs) besides conventional Internet applications running on laptop computers or enhanced PDAs will dominate. Then Video and Audio Streaming applications [52] and *Large Data Transfer* (LDT) applications including the *Multimedia Messaging Service* (MMS) based on WAP version 2.0 [121] are felt to become more popular with the optimization of GPRS and with the introduction of EDGE and the related packet data service *Enhanced General Packet Radio Service* (EGPRS). With the evolution of EDGE to GERAN and the convergence with UTRAN towards the IP-Multimedia Subsystem defined in 3GPP Release 5 (see Section 5.2.3.2), *Voice over IP* (VoIP) technology will be introduced in the radio network and VoIP applications are expected to smoothly replace circuit-switched voice services.

This chapter gives a general survey of the different emerging mobile data applications and will present related traffic models. The traffic models are based on well-known models in the literature modified for the use for cellular packet radio networks traffic engineering. Frequently used applications like e-mail, WWW, WAP or FTP are outlined as well as emerging applications like Video and Audio Streaming and Voice over IP.

Traffic traces and packet size histograms are provided, gained by stochastic simulation of the traffic models over the TCP/IP stack and a narrowband channel model for the transmission of the generated IP packets defined by a fixed data rate, an Internet packet delay and a packet loss probability.

6.1 WWW Applications

The Internet represents the largest existing *Wide Area Network* (WAN) offering a broad spectrum of information resources in form of technical reports, product information, software, images, audio and video sources and electronic commerce services. The popularity among larger user groups to use these resources was driven by the introduction of the *World Wide Web* (WWW). In the WWW, global naming conventions, protocols and object formats are used. An object can be a text file, an audio source, an image file or even the code of a program that can be interpreted and executed on the user's terminal (Java applet). *Graphical User Interfaces* (GUIs), so-called *WWW browsers*, enable the user to navigate within the WWW and to transfer and display WWW objects. Besides the use of the WWW as the public Internet, the WWW concept is increasingly introduced in company networks. By the introduction

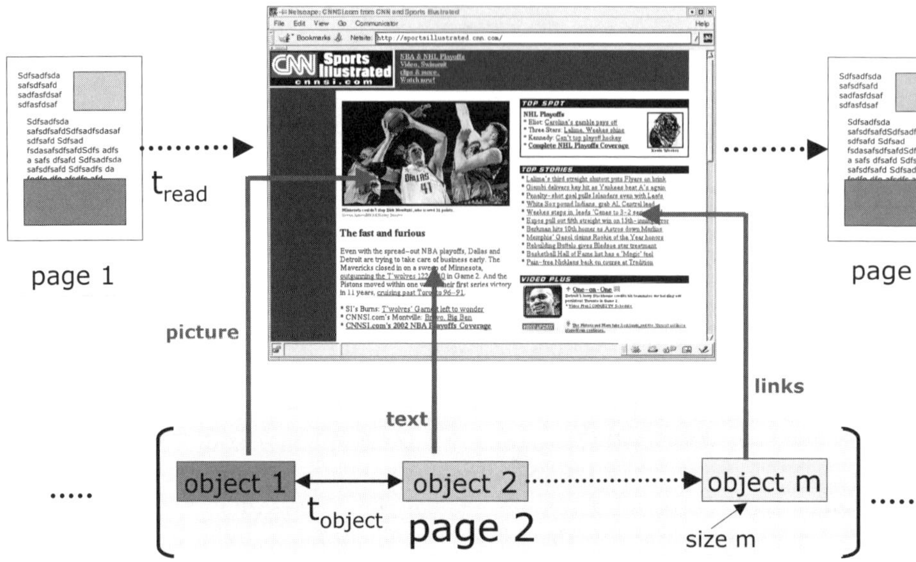

Figure 6.1: Structure of a WWW page

of such *intranets* a higher transparency of company processes and the simple usage of company databases as well as the optimized management of customer relationship (CRM) is expected and realized.

A WWW browser displays WWW pages that are composed of one or more WWW objects. Figure 6.1 shows a WWW page displayed by a WWW browser. It contains several different objects such as text and images. Within one page *links* can be defined that are connected to an identifier of a WWW resource, a *Unified Resource Locator* (URL). When the user chooses a link, a new object or page is displayed. By the direct specification of a URL in the text field of the WWW browser, a specific page or object can be chosen by the user (see Figure 6.1). Every object of the WWW has a unique URL.

The presentation of a WWW page is described by the *Hypertext Markup Language* (HTML). In an HTML file the appearance and the URLs of the contained objects are specified. An HTML page can also contain text that is displayed at the related position of the page. An HTML page itself is also a WWW object that is analyzed by the browser. Further WWW objects are specified in the HTML file as parts of the page that are requested and transmitted.

The WWW is characterized by a client-server architecture. A WWW browser represents a client that poses requests to a WWW server concerning the transmission of WWW objects that are stored on the server. The communication between client and server is controlled by the *Hypertext Transfer Protocol* (HTTP) [56]. HTTP is an application-oriented protocol that uses the *Transport Control Protocol* (TCP) [95] for end-to-end transport. In the initial version of HTTP a new TCP connection is set up for each requested object. The client transmits a request for a certain object to the server. The request contains the URL of the object, data format parameters and parameters for access control. The server processes this request and transmits the requested object to the client. Then the TCP connection is released. In newer HTTP

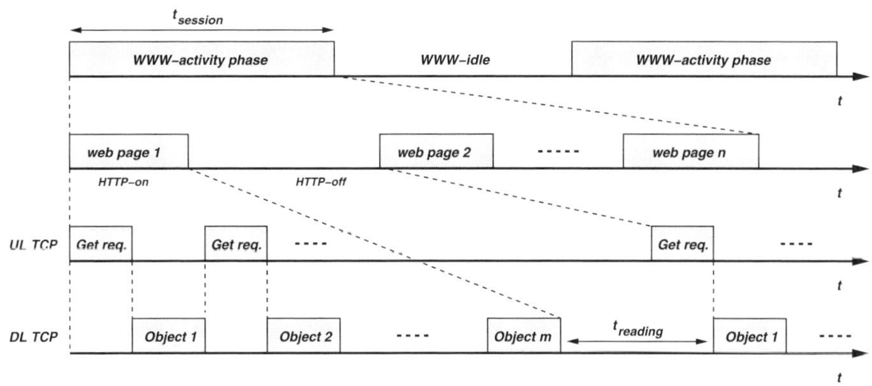

Figure 6.2: Sequence and parameters for a WWW session

Table 6.1: Mosaic WWW traffic model parameters

WWW Parameter	Distribution	Mean	Variance
Pages per session	geometric	5.0	20.0
Intervals between pages [s]	negative exponential	12.0	144.0
Objects per page	geometric	2.5	3.75
Object size [byte]	$\log_2$-Erlang-k ($k = 17$)	3700	$4.67 \cdot 10^9$
	transformed Erlang	9.4	5.2

versions, e.g., HTTP version 1.1, TCP connections can be reused by the following objects (persistent TCP connections) and several TCP connections can be set up in parallel to be used for the pipelined transmission of several different objects (pipelined TCP connections). Figure 6.2 depicts the sequence of a WWW session.

6.1.1 Adapted Mosaic WWW Model

As mentioned above WWW sessions consist of requests for a number of pages. These pages consist of a number of objects with a certain object size. Another characteristic parameter is the delay between two pages depending on the user's behavior to surf around the Web (see [33, 54]). Table 6.1 gives an overview of the Mosaic WWW traffic parameters. The small number of objects per page (2.5 objects), and the small object size (3700 byte) were adapted based on [54], since Web pages with a large number of objects or large objects are not suitable for thin clients such as PDAs or smart phones served by (E)GPRS. The traffic characteristics of the WWW model can be seen in the distribution function of the object size.

In Figure 6.3 histograms for the object size and the session duration are shown. They were gained during stochastic simulation of the traffic models over the TCP/IP stack and a narrowband channel model for the transmission of the generated IP packets with a fixed data rate of 89.6 kbit/s (typical user data rate for EDGE/UMTS), an IP packet delay of 110 ms and without packet loss in the Internet.

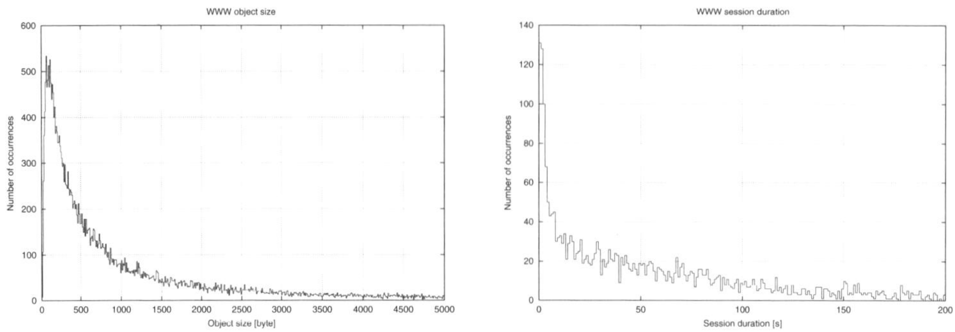

(a) Object size (b) Session duration

Figure 6.3: WWW traffic characteristics

6.1.2 Choi's Behavioral Model of Web Traffic

Choi's behavioral model is newer than the Mosaic model and therefore is representative for WWW browsing performed on *Personal Computers* (PCs) connected to the fixed Internet. Therefore the model is characterized by higher object sizes and larger WWW pages than in the Mosaic model. Since the performance of mobile networks in the first years of the GSM evolution is lower than in fixed access networks, this model is less suitable for GPRS traffic engineering. Once the performance of evolved GSM networks will be comparable to the performance of today's fixed access networks (e.g. with EGPRS or UMTS), the importance of this model will increase for cellular packet radio networks traffic engineering problems.

Choi's traffic model is an on/off-model with alternating phases of packet generation and silence (see Figure 6.4) [43]. An on-phase starts after the arrival and acceptance of a web request. During this phase, the objects of a WWW page are requested. The off-phase represents a silence period after all objects have been retrieved. Thus, the on and off-phases equal the page loading times and page viewing times, respectively.

During the on-phase the page's objects are downloaded. We distinguish two types of objects: the main object containing the document's HTML code and inline objects, such as linked objects, images or Java applets. To fetch all those inline objects modern browsers open several TCP connections in parallel after the successful retrieval of the main object. In Table 6.2 the random variables used by Choi to describe the object sizes, the number of inline objects and the length of the viewing time are listed.

Table 6.2: Choi WWW traffic model parameters

WWW Parameter	Distribution	Mean	Variance
Viewing time [s]	Weibull	39.5	$8.57 \cdot 10^3$
No. of inline objects per page	Gamma	5.55	130.0
Size of main object [kB]	Log-Normal	10	625.0
Size of inline objects [kB]	Log-Normal	7.7	$1.59 \cdot 10^4$

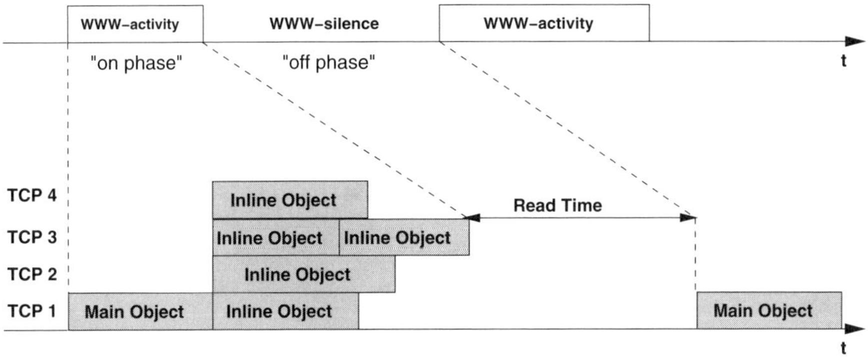

Figure 6.4: Choi's behavioral model

The amount of data S that is generated during the on-phase is given by

$$S = s_{Main} + \sum_{i=0}^{n_{Inline}} s_{Inline,i} \tag{6.1}$$

which is a combination of three random variables. The random variable S represents the payload to be transmitted by the HTTP protocol.

6.2 E-mail Applications

E-mails are transmitted in the Internet by using the *Simple Mail Transfer Protocol* (SMTP) [96] or the protocols *Post Office Protocol version 3* (POP3) [89] and *Internet Message Access Protocol* (IMAP) [45] for e-mail download.

6.2.1 SMTP, POP3 and IMAP

While local mail delivery systems, e.g., in *Local Area Networks* (LANs), are often based on the SMTP, it is often not possible to keep an e-mail client interconnected to the Internet for a long amount of time. Therefore the protocols POP3 and IMAP are widely used to download e-mail from a so-called mail-drop service (see Figure 6.5). These protocols are deployed to permit a terminal to retrieve mail that the server is holding for it, called the *store-and-forward* concept. TCP is used for e-mail transfer as the end-to-end transport protocol.

6.2.2 E-mail Traffic Model

Since the size of an e-mail download on a mobile device is the crucial parameter for cellular packet radio networks traffic engineering, a traffic model defining e-mail sizes is suitable. The introduced e-mail model based on [92] describes the load arising with the transfer of messages performed by a user. The only parameter is the e-mail size that is characterized by two $\log_2$-normal distributions plus an additional fixed quota of 300 byte (see Table 6.3). The base quota was assumed to be a fixed overhead. Subtracting the overhead, a bimodal distribution remained. The lower 80% can be interpreted as text-based mails, while the upper 20% could represent

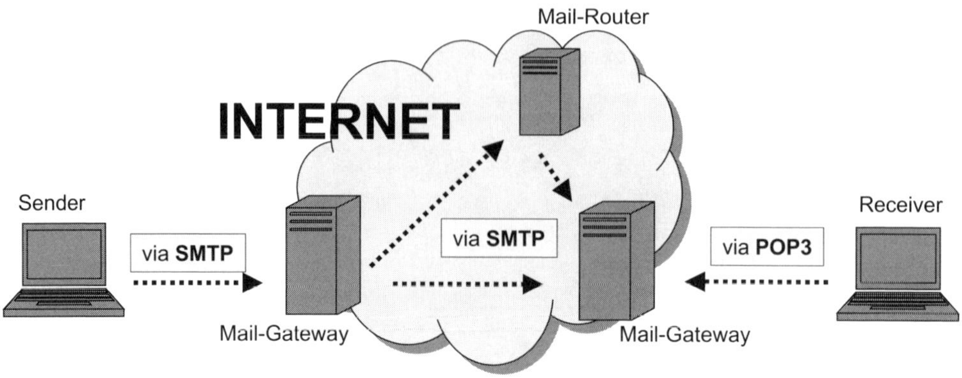

Figure 6.5: SMTP and POP3 for e-mail transfer

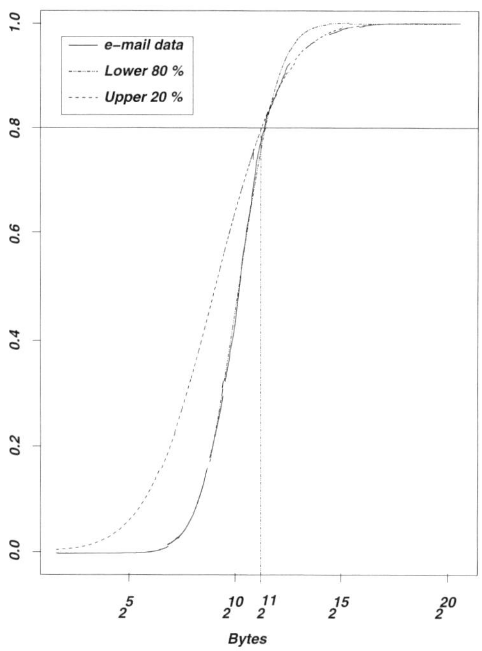

Figure 6.6: Bimodal $\log_2$ normal distribution function

mails with attached files, which can be rather large. The transition between these two distributions is 2 kbyte (see Figure 6.6). The maximum e-mail size is set to 100 kbyte.

In Figure 6.7 the histograms for the lower 80% of small e-mails and for the total range of the e-mail size are presented. Two different classes of e-mails can be observed. The size of the text-based e-mails begins with 300 byte, which is the default size needed for header information. The borderline between the distribution functions of these two e-mail classes is 2 kbyte according to [92].

Table 6.3: E-mail traffic model parameters

E-mail Parameter	Distribution	Mean	Variance
E-mail size (lower 80%) [byte]	$\log_2$-normal	1700	$5.5 \cdot 10^6$
	transformed normal	10.0	2.13
E-mail size (upper 20%) [byte]	$\log_2$-normal	16000	$71.3 \cdot 10^9$
	transformed normal	9.5	12.8
Base quota [byte]	constant	300	0

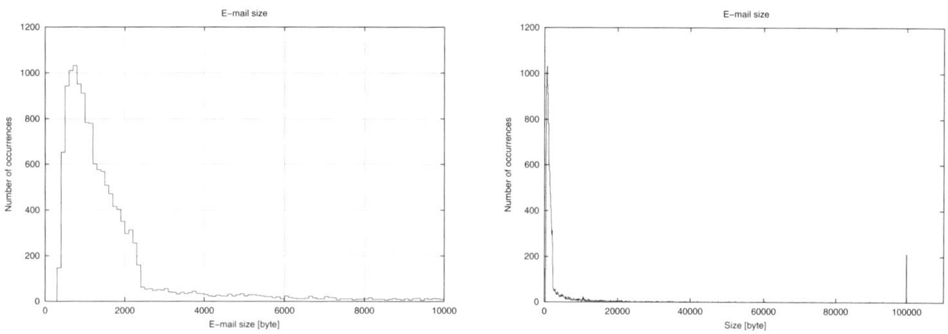

(a) Small sizes (b) Total range

Figure 6.7: E-mail size

6.3 File Transfer Applications

The *File Transfer Protocol* (FTP) [97] was developed to enable the safe, secure and efficient file transfer between two computers. System-specific features like the structure of file systems of the computers involved in the transfer are overcome with this standard. The term FTP is commonly used both for the protocol and for the file transfer application program.

6.3.1 FTP Protocol

FTP is an application-oriented protocol that uses TCP as the transport layer. The systems communicating over FTP are building a client-server relationship. A system is called an FTP client when it poses requests to another system, the FTP server, concerning the transfer of files. The specified file is then transferred between client and server.

FTP uses two connections between client and server, which are called *FTP-Control* and *FTP-Data* (see Figure 6.8). An FTP control connection is used for communication control and command transfer managed by the *Protocol Interpreter* (PI). Examples of commands are `dir` for the listing of a directory or the commands `put` and `get` for the transmission of files between client and server. Only one control connection is used during an FTP session, while for each transmitted object one data connection is used managed by the *Data Transfer Process* (DTP). One object can be a file or, e.g., a directory listing.

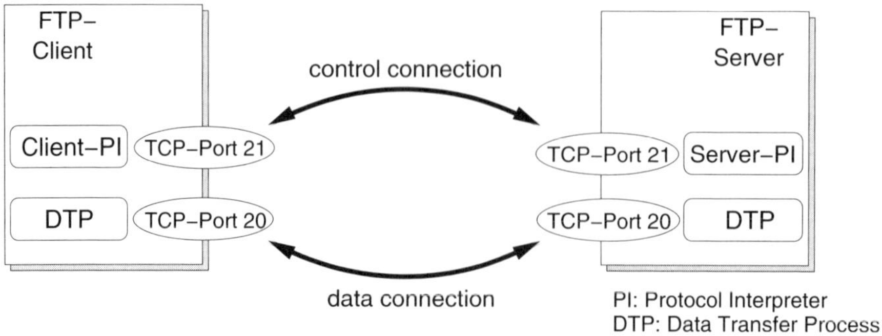

Figure 6.8: FTP data and control connection

Table 6.4: FTP traffic model parameters

FTP Parameter	Distribution	Mean	Variance
Total amount of data [byte]	$\log_2$-normal	32542	9661
	transformed normal	15.0	$1.9 \cdot 10^{-5}$
Amount of data per file [byte]	$\log_2$-normal	3000	1000
	transformed normal	11.55	$2.31 \cdot 10^{-4}$
Interval between connections [s]	$\log_{10}$-normal	4	2.54
	transformed normal	0.57	0.028

The importance of FTP is decreasing with the wide usage of HTTP downloads and e-mail attachments. Particularly in wireless networks the usage of FTP is expected to diminish.

6.3.2 FTP Traffic Model

The FTP model introduced in this section represents a unidirectional data transfer. It describes a file transmission from an FTP server to an FTP client. The examinations are based on extensive WAN measurements documented in [92]. Over 95% of the measured FTP data connections are performed by a `get` command. Since the downlink represents the bottleneck in cellular networks supporting Internet applications, which are asymmetric in nature, it is sufficient to regard only data transfer from server to client. Traffic originated by FTP control connections is not considered in the following model. Parameters describing an FTP session are:

- the total bulk of data per session
- the size of each transferred object
- the interval between two object transmissions

The bulk of data per session characterizes the duration of a session. A *log_2-normal distribution* is proposed for this total data per session and for the object size likewise. The given number of objects includes both the file transfer and, e.g., listing of directories. To describe the interval between two object transmissions a model created in

[93] can be used. Intervals between two objects should follow a *log_{10}-normal distribution.* All parameters, mean and variance of the distribution functions as well as the transformed mean and variance values are summarized in Table 6.4.

6.4 WAP-based Applications

The WAP specifications, which are implemented in today's mobile terminals, including WAP 1.2.1, aim at optimizing operation in 2G networks. Therefore WAP 1.2.1 defines a distinct technology comprising protocols and content representation. WAP is a suite of specifications that defines an architecture framework containing optimized protocols (e.g., WDP, WTP, WSP), a compact *Extensible Markup Language* (XML)-based content representation (*Wireless Markup Language* (WML), *WAP Binary Extensible Markup Language* (WBXML)) and other mobile-specific features like *Wireless Telephony Application* (WTA) [117, 118].

6.4.1 WAP Release 1.x

In addition to the goal of optimized operation in 2G networks, WAP had been developed because graphics-enhanced web services could not be brought to and displayed on thin clients, e.g., GSM mobile phones, and IP was not applicable in some environments, e.g., WAP over *Short Message Service* (SMS) or *Unstructured Supplementary Service Data* (USSD). Because of the optimization and the different protocols it was not possible to run WAP end-to-end to regular Internet sites. Instead, a WAP Gateway must be used. The main service a WAP Gateway provides is a protocol relay function between WAP stack and Internet stack. In addition to this standardized functionality, many gateway vendors provide a variety of value-added services that allow for personalization, for example.

The WAP protocol architecture (see Figure 6.9) comprises the following components:

Wireless Application Environment (WAE) With WAE an environment for development and execution of services is provided. The core component of WAE is the micro browser that renders the respective WML content.

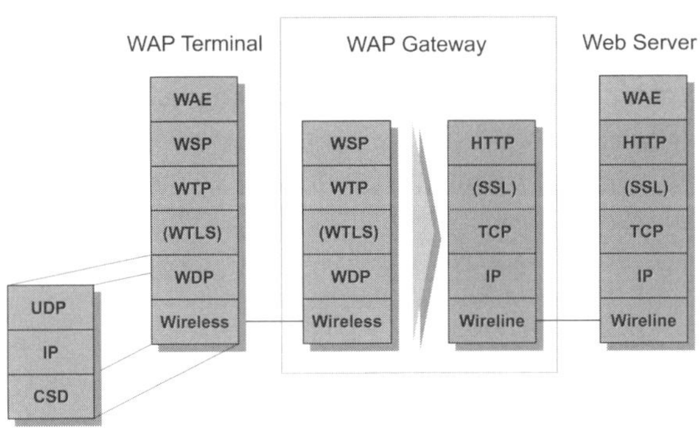

Figure 6.9: The WAP protocol architecture

Wireless Session Protocol (WSP) The session layer protocol WSP contributes methods to handle the session context. Thus communication between client and server is enabled.

Wireless Transaction Protocol (WTP) The transaction layer protocol WTP layer provides connectionless as well as connection-oriented communication. WTP class 2 represents the acknowledged client-to-server communication, which is commonly used for most of the WAP traffic.

Wireless Transport Layer Security (WTLS) WTLS is an optional function, which works in a similar way to *Secure Socket Layer* (SSL) or *Transport Layer Security* (TLS) in the Internet. WTLS provides authentication and privacy for ongoing communication.

Wireless Datagram Protocol (WDP) The transport layer WDP is an adaptation layer between WAP and the underlying bearer. In case of IP as the network layer, UDP is used instead of WDP.

Compared to the well-known Internet protocols like TCP and HTTP, WAP protocols differ in the following aspects:

- WDP or UDP provides an unacknowledged datagram transport only. This approach avoids overhead and delay for services that operate on a connectionless basis. One example is a WTP class 0 transaction.
- In-order delivery and retransmission are handled by a separate transport layer. It provides these functions only if they are requested by a higher-layer service (WTP class 1 or 2).
- WAP works in a transaction-oriented way. The amount of data that can be transferred by one transaction defaults to 1400 byte. It can be negotiated at WSP session setup time. If WTP segmentation and reassembly is used the amount of data per transaction is limited by the negotiated SDU size.
- WSP implements all features of HTTP 1.1 and enhances these by a binary encoding aiming at greater efficiency over the air.

6.4.2 WAP Release 2.0

In the specification WAP 2.0 [121, 120] some existing WAP protocols have been extended by new capabilities. WAP 2.0 converges with widely used Internet protocols like TCP and HTTP. *Internet Engineering Task Force* (IETF) work in the *Performance Implications of Link Characteristics* (PILC) Working Group has been leveraged to develop a mobile profile of TCP for wireless links. This profile is fully interoperable with the common TCP that operates over the Internet today.

The inclusion of *Wireless Profiled TCP* (WP-TCP) into the WAP architecture of release 2.0 has been motivated by the introduction of wireless networks with higher data rates (e.g. GPRS, EGPRS and UMTS). The advantage of TCP is its ability to transfer large data, end-to-end security (using *Transport Layer Security* (TLS)) and convergence with IETF protocols. WP-TCP is optimized for wireless communication systems, but can interoperate with standard TCP implementations.

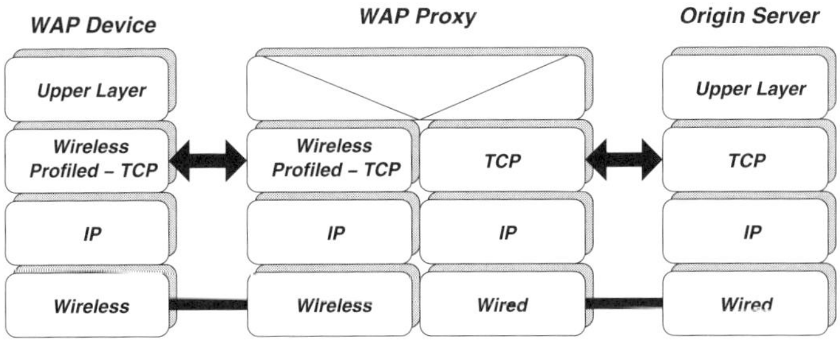

Figure 6.10: Wireless profiled TCP with WAP proxy

Further, WAP 2.0 does not require a WAP proxy, since the communication between the client and the server can be conducted using HTTP 1.1. However, deploying a WAP proxy can still optimize the communication process and may offer mobile service enhancements, such as location, privacy, and presence based services. In addition, a WAP proxy remains necessary to offer Push functionality.

Wireless Profiled TCP (WP-TCP) implementations can also be used for end-to-end connectivity without intermediate nodes. Thus WP-TCP must support both modes of operation, split- and end-to-end TCP. Which mode of TCP is chosen depends on factors such as the current provisioning, the application or the network access point. The requirements for this WP-TCP implementation are described in detail in [58]. It is generally based on standard TCP with additional features concerning TCP optimization for wireless links. An example list of additional options is as follows:

- Large Window Size based on Bandwidth Delay Product
- Window Scale Option
- Timestamp Option
- Large Initial Window
- Selective Acknowledgement Option (SACK)
- Path *Maximum Transfer Unit* (MTU) larger than default IP MTU
- Explicit Congestion Notification

In addition to protocol work, the WAP Forum has continued its work on service-enabling features for the mobile environment, like the push service or synchronization issues.

Although WAP 2.0 has been finished in 2001, WAP 1.x protocol stacks will still be used in the mobile terminals in the next years. For the WAP traffic model presented in the following, only WAP 1.2.1 is regarded.

6.4.3 Multimedia Messaging Service (MMS)

The *Multimedia Messaging Service* (MMS) is intended to provide a rich set of content, e.g., images, photographs in combination with voice or text, to subscribers in a messaging context. It supports both sending and receiving of such messages by

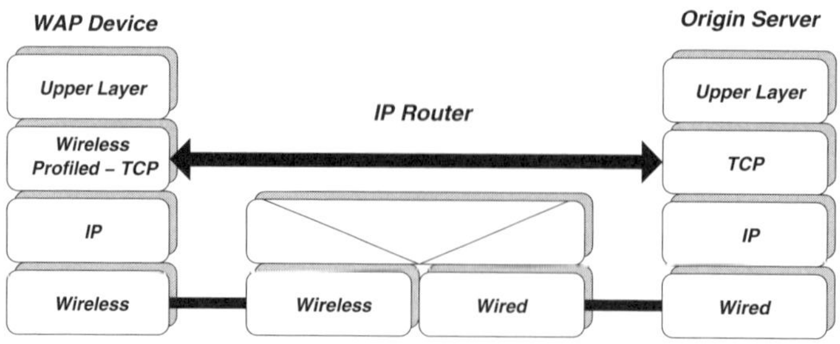

Figure 6.11: Wireless profiled TCP without WAP proxy

properly enabled client devices and is comparable to many messaging systems in use today (see Section 6.2). Examples include traditional e-mail available on the Internet and wireless messaging systems such as paging or SMS. These services provide a store-and-forward usage paradigm and it is expected that the MMS will be able to interoperate with such systems [119].

MMS notifications inform the MS about a received message. The MMS notification message consists only of MMS headers. The purpose of the notification is to allow the client to directly fetch a multimedia message from the location indicated in the notification. A client retrieves messages by sending a WSP/HTTP `get` (see Section 6.4.1) request to the MMS server containing the URL of the received message. When successful, the response to the retrieve request contains headers and the body of the incoming message.

MMS presentation is done by means of ordering, layout, sequencing and timing of multimedia objects on the terminal screen and other devices such as a speaker. With MMS presentation, the sender of the multimedia message has the possibility of organizing the multimedia content in a meaningful order and instructing how the multimedia objects are rendered at the receiving terminal.

There are various alternatives for the presentation language, most notably WML (see Section 6.4.1) and *Synchronized Multimedia Integration Language* (SMIL) [110]. While the WML presentation for multimedia messaging offers the same sequencing and layout capabilities as for browsing applications, the SMIL provides extended capabilities, such as timing of multimedia objects as well as animation. The SMIL is a simple XML-based language that consists of a set of modules that define the semantics and syntax for certain areas of functionality. Examples of these modules are the layout module, the timing and synchronization module and the animation module.

The MMS presentation language information is transferred in the same message that is carrying the multimedia objects. Thus, a multimedia message is a compact package of multimedia objects and optional presentation information.

A realistic traffic model for this application does not exist, yet, because MMS is not introduced in existing radio networks. Since several components of an MMS correspond to different applications like e-mail or WWW browsing, existing traffic models could be applied for an MMS traffic model. Typical sizes of first multimedia messages are predicted to 20 kbyte and the maximum size because of memory restrictions is expected to be 100 kbyte.

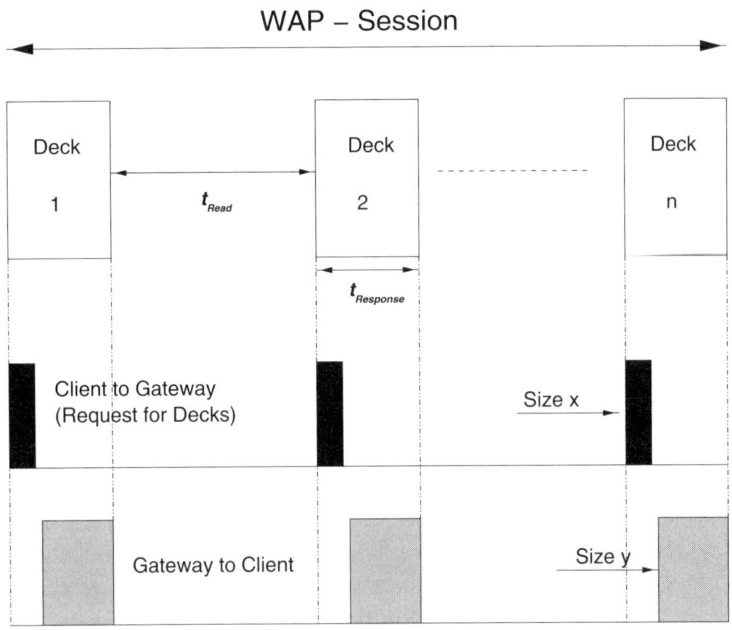

Figure 6.12: Sequence chart and important parameters for a WAP session

6.4.4 WAP Traffic Model

A WAP 1.2.1 traffic model was developed and applied in [105]. The main characteristics of the model are very small packet sizes (511 byte) approximately following a *log₂-normal distribution* and a limited value of the packet size (1400 byte). Newer WAP deck sizes are 1 kbyte for monochrome decks and 3 kbyte for colored decks.

A WAP session consists of several requests for a deck performed by the user. The sequence chart of a session is depicted in Figure 6.12.

WAP sessions are totally described by

- requests for a number of decks, n
- packet size in up- and downlink, $size\ x$ and $size\ y$
- a reading time the user takes before requesting the next deck, t_{Read}
- a response time of the network, $t_{Response}$

The latter is not determined by the user. This parameter depends totally on the underlying network.

Subsequent to [105], the number of decks is modeled by a *geometric distribution*, the reading time by a *negative exponential distribution* and the packet size by a truncated *log_2-normal distribution*. While the model of the session can be seen as general for WAP sessions, the parameters depend on the content. The parameters taken here are typical for WAP applications directly after introduction of WAP. Since the applications will change over the next years, the parameters of the model are subject to future updates. The parameters were derived by measurements performed at a WAP gateway in test operation.

Table 6.5: WAP 1.x traffic model parameters

WAP Parameter	Distribution	Mean	Variance
Decks per session	geometric	20.0	3800
Intervals between decks [s]	negative exponential	14.1	198.8
Size of 'Get Request' packet [byte]	$\log_2$-normal	108.2	$4.1 \cdot 10^3$
	transformed normal	6.34	0.71
Size of 'Content' packet [byte]	$\log_2$-normal	511.0	$3.63 \cdot 10^5$
	transformed normal	8.6	1.55

6.5 Streaming Applications

Many Internet portal sites are offering video services for accessing news and entertainment content from a Personal Computer (PC). Beside *Moving Pictures Expert Group* (MPEG), H.263 is the currently most accepted video coding standard for Video Streaming applications. In the area of Audio Streaming applications based on *MPEG Audio Layer-3* (MP3) are seen as potential applications. In the near future, mobile communication systems are expected to extend the scope of today's Internet Streaming solutions by introducing standardized Streaming applications as described in [52].

6.5.1 Video Streaming

6.5.1.1 H.263 Video Coder for Low Bandwidth

To transfer real-time video data generally high rates are necessary. Looking at an uncompressed video stream, even high capacity wireline networks hardly fit the needs of such a stream. Table 6.6 gives an impression of needed bit rates.

Without compression no manageable video transmission is possible. More or less all video codecs work on the same principles. They compress the original pictures by the reduction of redundant information and the removal of irrelevant data.

Redundancy reduction: the loss-free reduction of redundant information depends on signal properties and statistics. Therefore algorithms like predictive coding, coding with variable word length and run-level-coding are used.

Irrelevance removal: this lossy removal exploits the characteristics of human perception (e.g. fine details in an image can be omitted without visible distortion) by using different quantization levels.

Although video transmission at 64 kbit/s was already introduced in 1979 reasonable picture quality at bit-rates below 64 kbit/s is still difficult. However, in 1996 the

Table 6.6: Bit rates of common uncompressed video formats

Video format	Resolution	Bit rate
D1, digital studio standard	720 × 576 Pixel	166 Mbit/s
VHS quality (CIF)	352 × 288 Pixel	36.5 Mbit/s
Videophone (QCIF)	176 × 144 Pixel	3 Mbit/s

ITU-T Study Group XV specified a standard suitable for video transmission below 64 kbit/s. The international standard ITU-T Rec. H.263 [74] specifies a coded representation that can be used for compressing moving pictures for audio-visual services at low bit rates.

Another well-known and propagated video standard is the MPEG standard. H.263 offers great similarity to MPEG, since the experience with its use was included. It has MPEG-like blocks and macroblocks with prediction and motion compensation. The zigzagged quantized coefficients are coded using the MPEG run-level methods with different tables.

Overview of H.263 The basic configuration of this video source coding algorithm is based on ITU-T Rec. H.261 [71] and is a hybrid approach applying inter-picture prediction to utilize *temporal dependency* and transform coding of the remaining signal to remove *spatial redundancy*. Half pixel precision is used for the motion compensation, in contrast to Rec. H.261, where full pixel precision and a loop filter are used. Variable length coding is utilized for symbols to be transmitted.

In addition to the basic video source coding algorithm, negotiable coding options are included for improved performance. The most important ones are:

- *Unrestricted Motion Vector Mode* allows references which are outside the picture. For that purpose edge samples are duplicated to create the missing samples (H.263 annex D)
- *Syntax-based Arithmetic Coding* replaces the variable length coding for the same picture quality with fewer bits needed for coding (H.263 annex E)
- *Advanced Prediction Mode* sends a motion vector for each 8×8 luminance block for P frames (H.263 annex F)

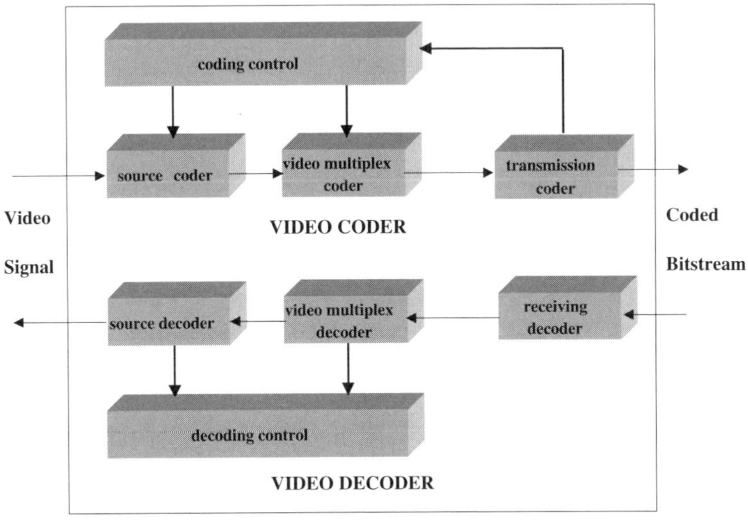

Figure 6.13: Outline block diagram of the H.263 video codec

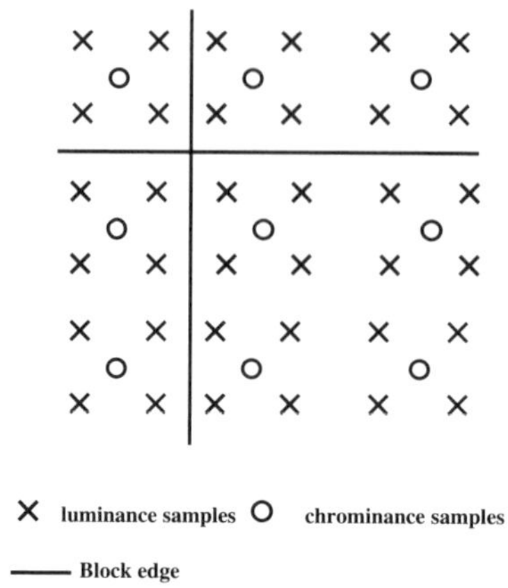

Figure 6.14: Sample position of luminance and chrominance

- *Bidirectional Predicted Frames* enable the generation of B-frames in between two successive P-frames, using both P-frames for prediction. Thus the remaining difference which has to be coded is minimized (H.263 annex G)

A block diagram of the H.263 codec is shown in Figure 6.13. The video coder consists of the source coder, the video multiplex coder and the transmission coder.

- The *source coder* reduces the spatial and temporal redundancies of the input video signal by employing a DCT (Discrete Cosine Transform) algorithm
- The *video multiplex coder* defines the hierarchical structure of the coded video data suitable for transmission
- Finally, the *transmission coder* controls the output bitstream to comply with the channel bandwidth requirements. It merges the data from other services (e.g. audio and data) and produces the coded bitstream, which is directed to the transmission channel

Source format To overcome the differences between the source formats like *Phase-Alternating Line* (PAL) and *National Television System Committee* (NTSC) specified for broadcast television systems, the ITU-T has adopted five standardized picture formats for H.263: sub-QCIF, QCIF, CIF, 4CIF and 16CIF. Here *CIF* stands for *Common Intermediate Format* (CIF) with the resolution of 352×288 pixels; Q-CIF is a quarter of CIF and 4CIF means four times the size of CIF.

Each pixel is represented by one luminance and two chrominance symbols. The chrominance signals are subsampled by a factor of two horizontally as well as vertically. Thereby the characteristics of the *Human Visual System* (HVS) regarding the response to colors are exploited. The color difference samples are arranged so that their block boundaries coincide with luminance block boundaries (see Figure 6.14).

The luminance sampling structure is dx pixels per line and dy lines per picture. dx/2 pixels of each chrominance component are placed horizontally and dy/2 lines per picture are arranged vertically. The values of dx, dy, dx/2 and dy/2 are given in Table 6.7 for each of the picture formats.

Table 6.7: H.263 picture formats

Picture Format	Number of luminance pixels (dx)	Number of luminance lines (dy)	Number of chrominance pixels (dx/2)	Number of chrominance lines (dy/2)
sub-QCIF	128	96	64	48
QCIF	176	144	88	72
CIF	352	288	176	144
4CIF	704	576	352	288
16CIF	1408	1152	704	576

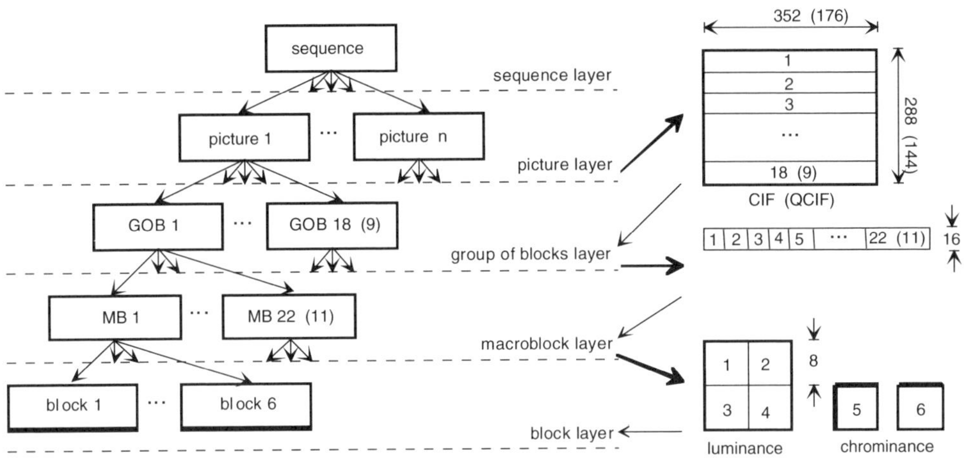

Figure 6.15: Hierarchical video frame structure

Hierarchical Structure of Video Frames Pictures are divided into a fixed number of lines called *Group Of Blocks* (GOB) depending on the format of the picture. A single GOB is divided again into a number of *Macro Blocks* (MBs) and finally these MBs are subdivided into *Blocks*.

The detailed structure, as an example outlined for CIF and QCIF, can be seen in Figure 6.15.

Table 6.8 lists the total number of GOBs, MB rows per GOB, MBs per row, and number of MBs in a GOB depending on the video format.

Format	GOBs	MB rows per GOBs	MB per row	MB in GOBs
sub-QCIF	6	1	8	8
QCIF	9	1	11	11
CIF	18	1	22	22
4CIF	18	2	44	88
16CIF	18	4	88	352

Table 6.8: Number of GOBs and MBs for different H.263 source picture formats

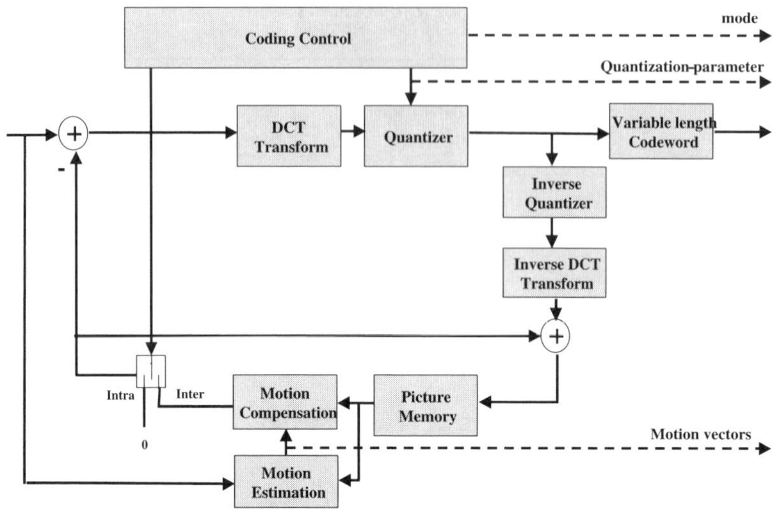

Figure 6.16: Block diagram of the H.263 source coder

H.263 Source Coder The source coder is the main body of the H.263 standard and specifies the picture format and the coding techniques used to achieve the bandwidth compression of the video data. The structure of the H.263 source coder is hybrid. On the one hand it utilizes interframe prediction exploiting temporal redundancy. On the other hand the coder applies transform coding of the residual prediction error exploiting spatial redundancy, and it adaptively reduces the spatial resolution. Temporal prediction is based on a block-based *Motion Estimation* and *Motion Compensation*, while a *Discrete Cosine Transform* (DCT) is used for spatial redundancy reduction.

Figure 6.16 shows the block diagram of the source coder. To support block tranformation incoming images are subdivided into non-overlapped blocks of 8×8 pixels. A block builds the smallest element in the hierarchy of video frames. The prediction error after motion compensation is transformed by the DCT. Depending on the values after tranformation the blocks are either transmitted or skipped. The coefficients of transmitted blocks are quantized and pairs of zero-runs and quantizer levels are combined, resulting in a *Variable Length Codeword* for each block. One of 31 uniform threshold quantizers may be selected adaptively. Finally motion vectors and quantization parameters are entropy coded along with other side information and multiplexed to the bitstream.

The coding configuration is also known as a hybrid *Differential Pulse-Code Modulation* (DPCM) transform coder. A simple DPCM loop is the generic structure of this configuration. The Picture Memory is included in the loop to memorize previously decoded frames. The main elements of the source coder are:

- Interframe prediction
- Block transformation
- Quantization

Prediction This coder is able to switch between two distinctive modes:

- **INTER**: pixel values are adopted from the prediction error provided by the frame difference between the original picture and the prediction of the DPCM loop
- **INTRA**: pixel values are taken directly from the original input sequence

P-frames (predictive frames), resulting from the INTER mode, are originated with reference to the most recent I-frame (intra frame) or P-frame. Therefore similar blocks of pixels common to two or more successive frames are replaced by a pointer that references one of the blocks. This so-called motion vector gives the displacement of the block in the current frame with reference to its match in the most preceding I- or P-frame, whichever is closest.

Frames from the INTRA mode are I-frames and are encoded independently of other frames. That means they contain the complete information of one picture and are therefore a lot larger than the P-frames. Thus they are used for at least the first frame, where no prediction is possible. I-frames can be generated in regular distances or adaptively depending on the loss situation. They operate as recovery points to restrict error propagation.

Motion Compensation Using the INTER mode the DPCM loop is active and prediction is mainly augmented by *motion compensation*. The decoder will accept one *motion vector* per macroblock or, if the *Advanced Prediction Mode* is used, one or four vectors per macroblock [74]. The prediction for the current MB is chosen so that either the correlation is maximized or the distortion is minimized. The horizontal and vertical displacements are stored in the motion vector and then transmitted as side information. Motion vectors are restricted so that all pixels referenced are within the coded picture area, unless if the Unrestricted Motion Vector mode and/or the Advanced Prediction mode is used.

DCT Transformation Only the transmitted blocks are processed by a separable two-dimensional *Discrete Cosine Transform* (DCT). A DCT is an orthogonal transform with optimized performance for natural scene images, which are often highly correlated. It is defined for a block of 8 × 8 pixels as follows:

$$F(u,v) = \frac{1}{8}C(u)C(v)\sum_{x=0}^{7}\sum_{y-0}^{7} f(x,y)cos\left(\frac{\pi(2x+1)u}{16}\right) cos\left(\frac{\pi(2y+1)v}{16}\right)$$

$$u,v,x,y = 0,1,2,...,7 \qquad C(u),C(v) = \begin{cases} \frac{1}{\sqrt{2}} & for \quad u,v=0 \\ 1 & otherwise \end{cases}$$

where x, y are the spatial coordinates in the pixel domain and u, v are the coordinates in the transformed domain.

Quantization and Clipping The DCT process does not achieve any compression. The DCT coefficients need further processing, which uses the advantage of typical image statistics. The frame difference signal tends to be made up of regions of similar values and only a few coefficients have significant magnitude at all. The quantization process sets all small values to zero and sets all non-zero-transformed values to the nearest preferred magnitude. 32 quantization step sizes exist. One is for the INTRA DCT coefficient and 31 for all other coefficients.

Syntax of Video Sequences in H.263 Figure 6.15 shows the syntax of H.263 video frames. The four-layered hierarchical structure corresponds to the subdivision of pictures. It consists of a picture layer and of sublayers for GOBs, MBs and blocks.

Within the picture layer multiple pictures can be coded. These pictures are further divided into GOBs and MBs. The MB always has a *Macro Block* (MB) header. Further on it is made up of six *blocks*; the first four are luminance blocks and the next two are chrominance blocks. For more information regarding the syntax see [74, 71].

6.5.1.2 Protocols for Real-Time Applications

Real-time Transport Protocol (RTP) RTP was introduced in 1996 by the IETF. It offers explicit support for multimedia traffic with needs for short latency but without paying attention to *Quality of Service* (QoS) issues [100].

RTP consists of two parts, one for the transport of data and the other for the transport of control messages. The latter is called *Real-Time Control Protocol* (RTCP). They are part of the Application Layer and independent from underlying transport protocols.

Transmitting multimedia data over wireless networks, e.g., audio- or video streams, packet loss up to a certain limit is acceptable. UDP has been chosen to transport the real-time data. Though it is not a reliable protocol, it is still very slim and fast in contrast to TCP, which in turn causes a lot of control traffic.

The original data flow from a server is blocked into consecutive packets. To be able to restore the chronological order, every packet in RTP is marked not only with an ascending sequence number, for example to detect packet loss, but also with a time stamp (see Figure 6.17). This allows the receiver of the packets to merge them again to a stream. Delay and jitter calculations are also simplified. Data arriving too late can be identified very quickly. The field for the Payload Type (PT) enables

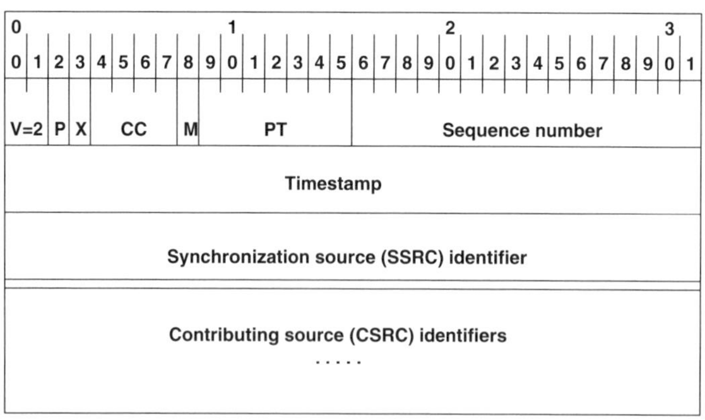

Figure 6.17: RTP header

the system to distinguish between different sorts of payload and to take care of their special needs. The header fields are defined in [124, 38].

Real-Time Control Protocol (RTCP) RTCP is based on the periodic transmission of control packets to all participants in a session. It is primarily dedicated to provide feedback on the quality of data distribution. Furthermore, it is used for the control of adaptive coders. Their packet format is similar to that of normal RTP packets but several types are distinguished:

- SR: Sender Report, statistics from participants that are active senders
- RR: Receiver Report, statistics from participants that are not active senders
- SDES: Source Description items
- BYE: indicates end of participation
- APP: application specific functions

Real-Time Streaming Protocol (RTSP) The RTSP is a protocol that is applied by the application layer. In contrast to RTP this protocol is not used for delivering a stream. It is a text based network remote control to handle the applications, which interoperates with the stream. With signals like `GET_PARAMETER`, `PLAY`, `RECORD` or `PAUSE` the protocol RTSP manages the status of sessions. It is similar to HTTP, which handles *Hypertext Markup Language* (HTML) documents in the WWW. RTSP expands its functionality according to the demands of streaming content. Details are given in [101].

6.5.1.3 Video Streaming Traffic Model

In the scope of modeling video sources, a lot of attention has been paid to long range dependent or self-similar models of traffic streams in telecommunication networks [116]. Many such models have been used to investigate *Variable Bit Rate* (VBR) video sources with a statistical analysis of empirical sequences and estimation of the grade of self-similarity [98]. Since MPEG and H.263 video traffic consists of a highly correlated sequence of images due to its encoding, the correct modeling of the correlation structure of the video streams is essential [123].

For the proposed traffic model in this work no stochastic models of video streams with self-similar or highly correlated traffic characteristics are applied. Real video sequences coded by an H.263 coder are used to generate the Streaming traffic.

Table 6.9: Offered IP traffic of video sequences

Sequences	Offered IP traffic	
	Q20	80-10-10 Mix
`Claire`	10.9 kbit/s	
`Carphone`	26.7 kbit/s	**14.39 kbit/s**
`Foreman`	31.7 kbit/s	

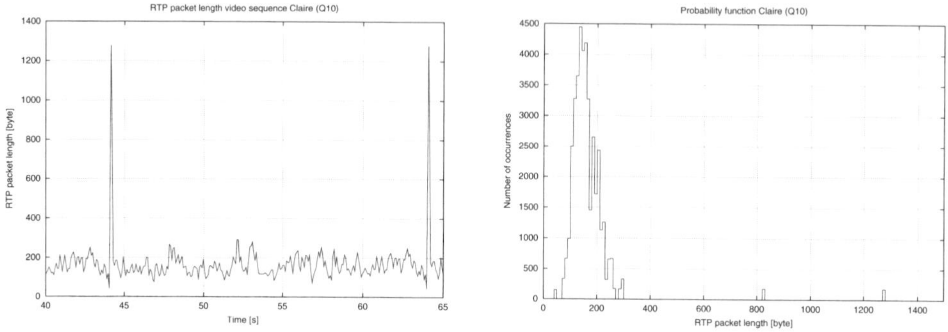

(a) Video trace (b) Probability function RTP packet length

Figure 6.18: Video sequence `Claire` (*Quantization Level 10* (Q10))

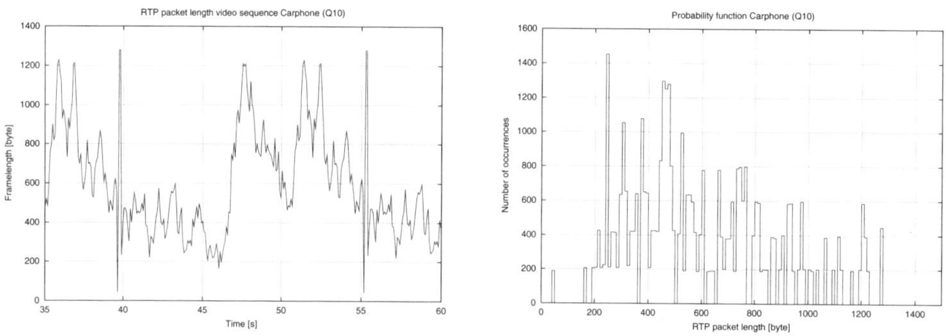

(a) Video trace (b) Probability function RTP packet length

Figure 6.19: Video sequence `Carphone` (Q10)

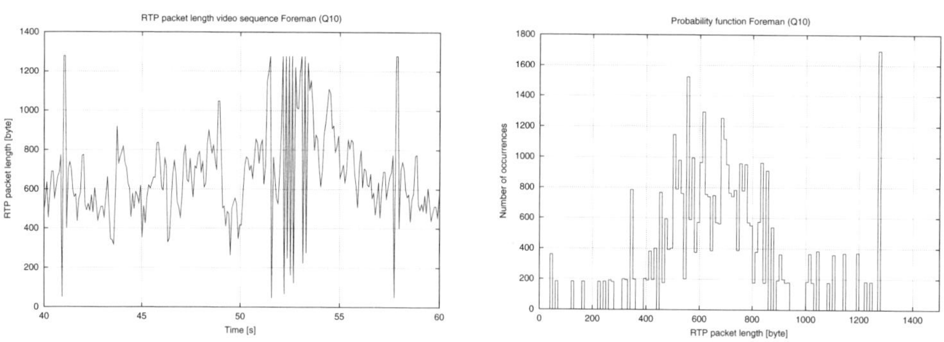

(a) Video trace (b) Probability function RTP packet length

Figure 6.20: Video sequence `Foreman` (Q10)

The Video Streaming traffic model used within the scope of this book is based on three video sequences in the *Quarter Common Intermediate Format* (QCIF) with the resolution of 176×144 pixels. The sequences are proposed by the *Video Quality Expert*

Group (VQEG) and are for this reason commonly used. Each sequence represents a particular group of videos with different intensities of motion.

- `Claire` stands for videos with a very low motion intensity and can be seen as a characteristic video conferencing sequence or inactive visual telephony.
- `Carphone` includes both periods with rather high motion and periods of low motion intensity. It represents many kinds of vivid or active video-conferences or even visual telephony.
- The third video, `Foreman`, is a sequence with permanently high motion intensity of both the actor and the background. This permanent motion is characteristic of sport events or movie trailers.

The H.263 coder was used with a skip factor of 2, which means that every second frame of the original sequence was skipped, so that the frame rate of the coded sequences was reduced from 25 to only 12.5 frames/s.

Q20 was adjusted for Intra (I)- and Predictive (P)-frames. The resulting video quality is low, but it is acceptable for mobile devices with their limited visual output capacities.

A conservative mix of sequences including 80% `Claire`, 10% `Carphone` and 10% `Foreman` has been selected for the proposed traffic model. The mix represents video streams with low motion and only a few streams with higher motion intensity.

Due to the negligible size of RTSP and RTP control messages in comparison to the size of real-time data, they are proposed to be neglected. The resulting average IP traffic offered by this particular mix is 14.39 kbit/s (see Table 6.9).

All of the new emerging applications are predicted to be relatively short in duration. So-called *heavy users*, generating long streams with huge amounts of data, have not been taken into account. The duration of video sessions is modeled by a negative-exponential distribution with an average value of 120 s. This is an assumption with regards to the prognosis for 3G networks in [54] where the duration of real-time calls is proposed to be modeled by a negative-exponential distribution.

In Figure 6.18, Figure 6.19 and Figure 6.20 the RTP packet length trace and histogram of the three video sequences `Claire`, `Carphone` and `Foreman` are presented.

6.5.2 Audio Streaming

6.5.2.1 MPEG-1/MPEG-2 Audio Layer-3: MP3

For a digital audio signal from a CD, the bit rate is 1411.2 kbit/s. With MP3 compression, CD-like sound quality is achieved at 128 kbit/s. Sound quality similar to FM radio for stereo signals can be achieved even at a bitstream of 64 kbit/s, and for a mono signal at only 32 kbit/s. For the use of low bit rate audio coding schemes in broadcast applications at bit rates of 60 kbit/s per audio channel, the International Telecommunication Union (ITU)-R recommends MPEG Layer-3.

Highest coding efficiency is achieved by algorithms exploiting signal redundancies and irrelevancies in frequency domain based on a model of human auditory systems.

All coders use the same basic structure (see Figure 6.21). The coding scheme can be described as perceptual noise shaping or perceptual subband/transform coding. The encoder analyzes the spectral components of the audio signal by calculating a filterbank (transform) and applies a psychoacoustics model to estimate the just

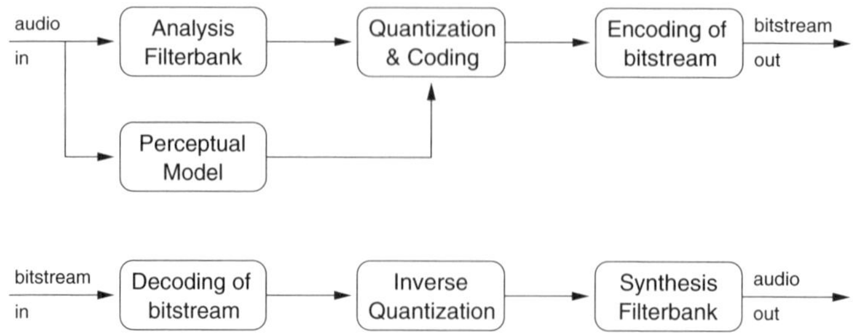

Figure 6.21: Block diagram of a perceptual encoding/decoding system

noticeable noise-level. In its quantization and coding stage, the encoder tries to allocate the available number of data bits in a way to meet both the bit rate and masking requirements.

It consists of the following building blocks:

- Filter bank: A filter bank is used to decompose the input signal into subsampled spectral components (time/frequency domain). Together with the corresponding filter bank in the decoder it forms an analysis/synthesis system.
- Perceptual model: Using either the time domain input signal and/or the output of the analysis filter bank, an estimate of the actual (time and frequency dependent) masking threshold is computed using rules known from psychoacoustics.
- Quantization and coding: The spectral components are quantized and coded with the aim of keeping the noise, which is introduced by quantizing, below the masking threshold. Depending on the algorithm, this step is done in very different ways, from simple block companding to analysis-by-synthesis systems using additional noiseless compression.
- Encoding of bitstream: A bitstream formatter is used to assemble the bitstream, which typically consists of the quantized and coded spectral coefficients and some side information, e.g. bit allocation information.

6.5.2.2 Psychoacoustics

The term psychoacoustics describes the characteristics of the human auditory system on which modern audio coding technology is based.

The sensitivity of human auditory systems for audio signals is one of its most significant characteristics. It varies in the frequency domain, e.g. the sensitivity is high for frequencies between 2.5 and 5 kHz and decreases beyond and below this frequency band. The sensitivity is represented by the *Threshold in Quiet* (see Figure 6.22). Any tone below this threshold will not be perceived.

The most important psychoacoustics fact is the masking effect of spectral sound elements in an audio signal like tones and noise. For every tone in the audio signal a masking threshold can be calculated. If another tone lies below this masking threshold, it will be masked by a louder tone and remains inaudible, too.

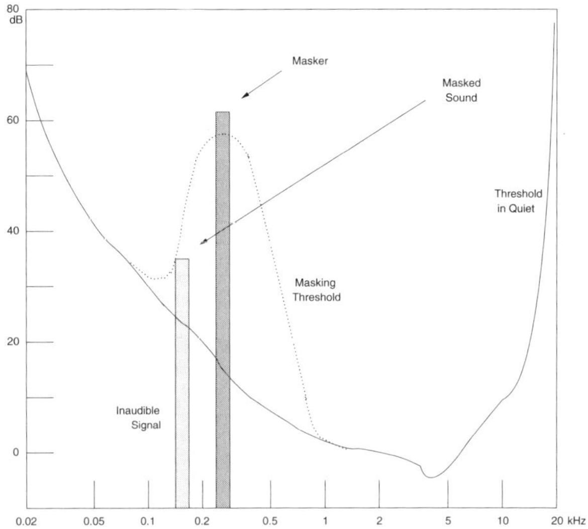

Figure 6.22: Psychoacoustics of human auditory system

These inaudible elements of an audio signal are irrelevant for human perception and thus can be eliminated by the encoder.

For the audio quality of a coded and decoded audio signal the quality of the psychoacoustics model used by an audio encoder is of prime importance.

6.5.2.3 Audio Streaming Traffic Model

Quite contrary to Video Streaming, the MP3 streams mainly consist of a *Constant Bit Rate* (CBR). In principle MPEG audio does not work just at a fixed compression ratio. The selection of the bit rate of the compressed audio is, within some limits, completely left to the implementer or operator of an MPEG audio coder. The standard defines a range of bit rates from 32 kbit/s (in the case of MPEG-1) or 8 kbit/s (in the case of MPEG-2 Low Sampling Frequencies extension (LSF)) up to 320 kbit/s (MPEG-1) or 160 kbit/s (LSF). In the case of MPEG-1/2 Layer-3, the switching of bit rates from audio frame to audio frame has to be supported by decoders. This, together with the bit reservoir technology, enables both variable bit rate coding and coding at any constant bit rate between the limits set by the standard. The Real-Audio (from RealNetworks Inc.) uses an adaptive streaming for alternating bandwidth. From most operators the CBR is chosen. For example, QuickTime streams (from Apple Inc.) and MP3 streams (offered by many Radio Stations) are realized using RTP.

Hence, the proposed traffic model for Audio Streaming over (E)GPRS is a CBR source with 10-64 kbit/s. For example the MPEG-2 *Advanced Audio Coding* (AAC) codec [69] generates 256 byte every 64 ms for 32 kbit/s Stereo and generating 128 bytes every 64 ms for 16 kbit/s Mono.

6.6 Voice over IP

The transfer of voice traffic over packet networks, and especially VoIP, is rapidly gaining acceptance. While many corporations have been using voice over *Circuit-Switched* (CS) networks to save money, the dominance of IP has increasingly shifted the attention from CS to *Packet-Switched* (PS) VoIP. VoIP transfer can significantly reduce the per-minute cost, resulting in reduced long-distance bills and a better utilization. In fact, many dial-around-calling schemes available today already rely on VoIP backbones to transfer voice, passing some of the cost savings to the customer.

But along with initial excitement, customers are worried over possible degradation in voice quality when voice is carried over these packet networks. Whether these concerns are based on experience with the early Internet telephony applications, or whether they are based on understanding the nature of packet networks, voice quality is a critical parameter in acceptance of VoIP services. Therefore, it is crucial to understand the factors affecting VoIP transmission, as well as to obtain the tools to measure and optimize them.

The key to providing a network architecture that allows new innovative services to be deployed quickly, efficiently and by a number of independent parties is to layer the network into three distinct functional areas: a Service Layer, a Call Layer and a Switching and Routing Layer. This is the vision that the *Session Initiation Protocol* (SIP) [65] provides for *Next Generation Networks* (NGNs) independent of whether fixed or mobile operators are involved. In fact, in some cases, it could be seen that when a SIP enabled, converged services core network is in place with an operator, the difference between a fixed and a mobile network simply becomes how a user accesses the service and applications delivered from the core. VoIP services need to be able to connect to traditional CS voice networks. The ITU-T has addressed this goal by defining H.323, the set of standards for packet-based multimedia networks [76], and the IETF by defining the SIP architecture (see Figure 6.23). For controlling Telephony Gateways from external call control elements called media gateway controllers or call agents, the *Media Gateway Control Protocol* (MGCP) [32] can be applied.

6.6.1 Session Initiation Protocol (SIP)

SIP is an application layer signaling protocol that defines initiation, modification and termination of interactive, multimedia communication sessions between users.

SIP is part of the IETF standards process and is modeled upon other Internet protocols such as SMTP and HTTP. It is used to establish, change and tear down (end) calls between one or more users in an IP-based network.

In order to provide telephony services there is a need for a number of different standards and protocols to come together - specifically to ensure transport (RTP), signaling inter-working with today's telephony network, to be able to guarantee voice quality, to be able to provide directories, to authenticate users, and to realize scalability for large subscriber numbers.

In essence, SIP has to provide the following functions:

- **Name Translation and User Location** - Ensuring that the call reaches the called party wherever they are located. Carrying out any mapping of descriptive information to location information. Ensuring that details of the nature of the call (Session) are supported.

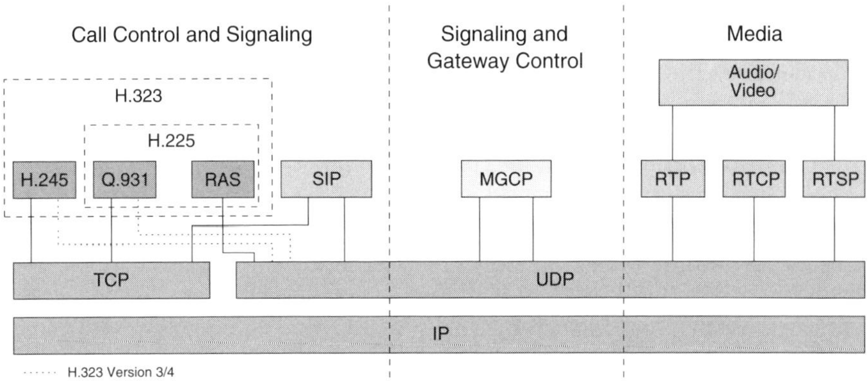

Figure 6.23: SIP, H.323 and MGCP

- **Feature Negotiation** - This allows the group involved in a call (this may be a multi-party call) to agree on the features supported - recognizing that not all the parties can support the same level of features. For example video may or may not be supported; as any form of *Multipurpose Internet Mail Extensions* (MIME) type is supported by SIP, there is plenty of scope for negotiation.
- **Call Participant Management** - During a call a participant can bring other users onto the call or cancel connections to other users. In addition, users could be transferred or placed on hold.
- **Call feature changes** - A user should be able to change the call characteristics during the course of the call. For example, a call may have been set up as voice only, but in the course of the call, the users may need to enable a video function. A third party joining a call may require different features to be enabled in order to participate in the call.
- **Protocol Components** - There are two components within SIP: the SIP User Agent and the SIP Network Server. The User Agent is effectively the end system component for the call and the SIP Server is the network device that handles the signaling associated with multiple calls.

6.6.2 Audio Codecs

Voice channels occupy 64 kbit/s using *Pulse Code Modulation* (PCM) coding when carried over CS links. Compression techniques (Codecs) were developed allowing a reduction in the required bandwidth while preserving voice quality.

6.6.2.1 Speech Codec (G.723)

The G.723 [73] is used for speech transmission in videoconferencing systems and is used in the video conference standard H.323.

The coder has two possible bit rates:

- 5.3 kbit/s

- 6.3 kbit/s

It encodes speech or other audio signals in frames using linear predictive analysis by synthesis coding. The excitation signal for the high rate coder is *Multi-Pulse Maximum Likelihood Quantization* (MP-MLQ) and for the low rate coder *Algebraic Codebook Excited Linear Predictive* (ACELP). The recommendation foresees a frame size of 30 ms in which the speech or other audio signals are encoded. Both rates are a mandatory part of the encoder and decoder. It is possible to switch between the two rates at any 30 ms frame boundary. An option for variable rate operation using discontinuous transmission and noise fill during non-speech intervals is also possible. In addition, there is a look ahead of 7.5 ms which gives a coding delay of 37.5 ms. All additional delays in the implementation and operation of this coder are due to:

- actual time spent processing the data in the encoder and decoder
- transmission time on the communication link
- additional buffering delay for the multiplexing protocol

The output of the speech coder during talkspurt and silence periods is depicted in Figure 6.24, generating speech packets of 24 byte at high rate and 20 byte when transmitting at low rate. Silence compression techniques reduce the transmitted data volume during silence intervals. Those techniques will be described in the following as *Voice Activity Detection* (VAD) and *Comfort Noise Generator* (CNG).

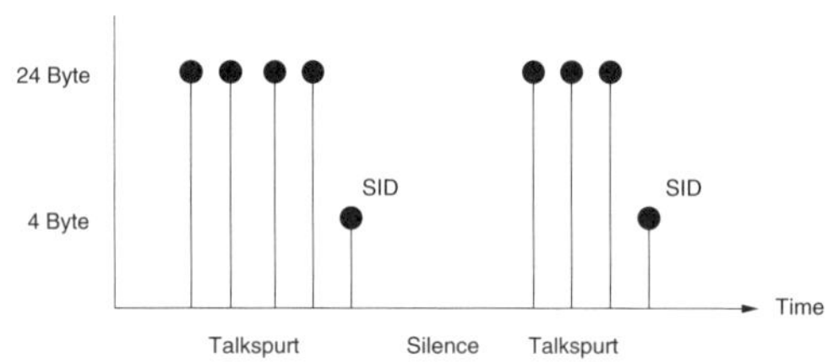

Figure 6.24: Output of speech coder G.723 - high rate (6.3 kbit/s)

6.6.2.2 Silence Compression Scheme

Silence compression techniques are used to reduce the transmitted bit rate during silent intervals of speech. Systems allowing discontinuous transmission are based on a VAD algorithm and a CNG algorithm that allow the insertion of an artificial noise during silence periods. This feature is necessary to avoid noise modulation introduced when the transmission is switched off: if the background acoustic noise that was present during active periods abruptly disappears, this very unpleasant noise modulation may even reduce the intelligibility of the speech.

Voice Activity Detection (VAD) The purpose of the VAD is to reliably detect the presence or absence of speech and to convey this information to the CNG algorithm. Typically, VAD algorithms base their decisions on several successive frames of information in order to make them more reliable and to avoid producing intermittent decisions. The VAD is constrained to operate on the same 30 ms speech frames which will subsequently be either encoded by the speech coder or filled with comfort noise by the comfort noise generator. The output of the VAD algorithm is passed to the CNG algorithm.

The largest difficulty in the detection of speech is the presence of any of a diverse range of background noise conditions. The VAD must be able to detect speech even in very low signal-to-noise ratio conditions. It is impossible to distinguish between speech and noise using simple level detection techniques when parts of the speech utterance are buried below the noise. The distinction between these conditions can only be made by taking into consideration the spectral characteristics of the input signal. In order to do this, the VAD incorporates an inverse filter, the coefficients of which are derived during noise-only periods by the CNG.

Comfort Noise Generator (CNG) The purpose of the CNG algorithm is to create a noise that matches the actual background noise with a global transmission cost as low as possible. At the transmitting end, the CNG algorithm uses the activity information given by the VAD for each frame, then computes the encoded parameters needed to synthesize the artificial noise at the receiving end. These encoded parameters compose the *Silence Insertion Descriptor* (SID) frames, which require fewer bits than the active speech frames and are transmitted during inactive periods.

The main feature of this CNG algorithm is that the transmission of SID frames is not periodic: for each inactive frame, the algorithm makes the decision of sending a SID frame or not, based on a comparison between the current inactive frame and the preceding SID frame. In this way, the transmission of the SID frames is limited to the frames where the power spectrum of the noise has changed.

During inactive frames, the comfort noise is synthesized at the decoder by introducing a pseudo-white excitation into the short-term synthesis filter. The parameters used to characterize the comfort noise are the *Linear Predictive Coding* (LPC) synthesis filter coefficients and the energy of the excitation signal. At the encoder, for each SID frame the algorithm computes a set of LPC parameters and quantizes the corresponding LSPs using the coder *Linear Spectral Pair* (LSP) quantizer on 24 bit. It also evaluates the excitation energy and quantizes it with 6 bit. This yields encoded SID frames of 4 byte including the 2 bit for bit rate and *Discontinuous Transmission* (DTX) information.

A notable feature of this CNG algorithm is the method used to evaluate the spectrum of the ambient noise for each SID frame. It takes into account the local stationarity or non-stationarity of the input signal.

6.6.3 Performance Issues

6.6.3.1 Delay

Delay causes two problems - echo and talker overlap. Echo is caused by the signal reflections of the speaker's voice from far-end telephone equipment back into the speaker's ear. Echo becomes a significant problem when the round-trip delay becomes greater than 50 ms. Since echo is perceived as a significant quality problem, Voice over Packet systems must address the need for echo control and implement some means of echo cancellation.

Talker overlap (or the problem of one talker stepping on the other talker's speech) becomes significant if the one-way delay becomes greater than 250 ms. The end-to-end delay budget is therefore the major constraint and driving requirement for reducing delay through a packet network.

Accumulation Delay: This delay (sometimes called algorithmic delay) is caused by the need to collect a frame of voice samples to be processed by the voice coder. It is related to the type of voice coder used and varies from a single sample time (0.125 ms) to many milliseconds, e.g., 30 ms for the G.723 codec (see Section 6.6.2.1).

Processing Delay: This delay is caused by the actual process of encoding and collecting the encoded samples into a packet for transmission over the packet network. The encoding delay is a function of both the processor execution time and the type of algorithm used. Often, multiple voice coder frames will be collected in a single packet to reduce the packet overhead.

Network Delay: This delay is caused by the physical medium and protocols used to transmit the voice data, and by the buffers used to remove packet jitter on the receive side. Network delay is a function of the capacity of the links in the network and the processing that occurs as the packets transit the network. The jitter buffers add delay which is used to remove the packet delay variation that each packet is subjected to as it transits the packet network. This delay can be a significant part of the overall delay since packet delay variations can be as high as 70-100 ms in IP networks.

6.6.3.2 Jitter

The delay problem is compounded by the need to remove jitter, a variable inter-packet timing caused by the network a packet traverses. Removing jitter requires collecting packets and holding them long enough to allow the slowest packets to arrive in time to be played in the correct sequence. This causes additional delay.

The two conflicting goals of minimizing delay and removing jitter have engendered various schemes to adapt the jitter buffer size to match the time varying requirements of network jitter removal. This adaptation has the explicit goal of minimizing the size and delay of the jitter buffer, while at the same time preventing buffer underflow caused by jitter.

6.6.3.3 Header compression

One of the problems with IP over wireless is the large overheads introduced by IP and other above protocol headers such as UDP and RTP (see Figure 6.25).

These headers are used to transmit each packet to the correct host with the proper application in the correct order and at the right time. In case of real-time IP services, IP/UDP/RTP protocol stack is used to carry the media frames. When transporting packet voice frames, the length of the RTP/UDP/IP header (40 byte for IPv4 and 60 byte for IPv6) is larger than the payload. Namely, GSM codecs use 20 ms frames. This corresponds to 32.5 bytes every 20 ms in the case of the GSM Full Rate codec. The overhead problem is addressed by header compression so as to enable VoIP to be transmitted in a spectrum efficient way. Header compression algorithms must be

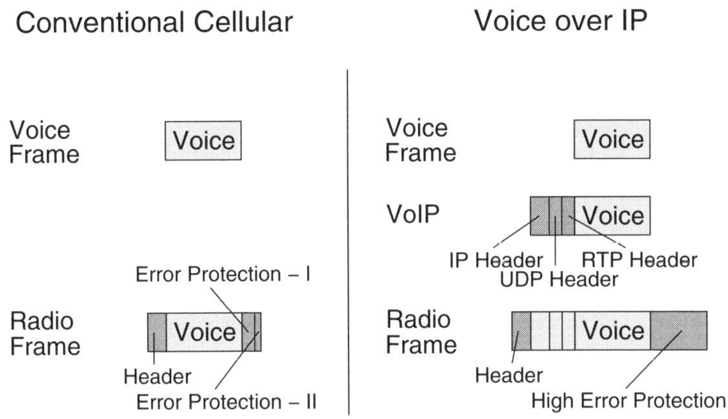

Figure 6.25: Voice frame

efficient and robust against errors to be used on the air interface. The proposed *Compressed RTP* (CRTP) algorithm, which is used for wireline VoIP, is not robust enough for wireless links and more relevant algorithms have been proposed in the ROHC specification [37].

6.6.4 Voice Traffic Characteristics

Voice activities can be considered as alternating between two states: talkspurt and silent. Data are generated during talkspurt only, and no data is transmitted during silence, thereby making statistical multiplex gain possible. Paul T. Brady discovered that both talkspurt and silence periods of digitized voice are exponentially distributed [41].

6.6.4.1 Six-state Model

Figure 6.26 shows the model for two speakers developed in [41]. Allowable state transitions are indicated in Figure 6.26. There is no attempt to control *B's* behavior; he starts and stops talking in his own manner. Notice that his state changes cause horizontal transitions.

Vertical transitions are determined by A. If he is talking, he stops when an *all silent pulse* occurs, and if silent he starts when a *start talking pulse* occurs. We call these β- and α-pulses, respectively. These pulses are a result of Poisson processes, so that, for example, if A is talking and B is silent (A solitary talk state), he stops talking in the next dt seconds with probability $\beta^A_{sol} \cdot dt$. For notation, the subscript on β, the fall silent parameter, describes the present state, while the subscript on α, the start talking parameter, denotes the event that will occur if the pulse occurs. The superscript refers to A or B. The six values for β and α are denoted $\beta_{sol}, \beta_{ted}, \beta_{tor}, \alpha_{pse}, \alpha_{alt}$, and α_{int} (see Figure 6.26) in which the abbreviations mean *sol*itary, interrup*ted*, interrup*tor*, pau*se*, *alt*ernate, and *int*errupt.

6.6.4.2 Single Speaker Model

The commonly accepted model for a speaker in a voice call is a continuous-time, discrete-state Markov chain. The holding time in each state is assumed to be exponentially distributed with mean $1/\lambda$ and $1/\mu$, respectively (see Figure 6.27).

The commonly used values are $1/\lambda = 0.65\,s$ and $1/\mu = 0.352\,s$. These values are often referred to as the *Traditional Packet Voice Model.* $1/\lambda = 0.7428\,s$ and $1/\mu = 0.4357\,s$ were also measured from actual telephone traffic with silence detector of high sensitivity [122]. The fundamental basis of all packet voice performance studies for the past is the classic model of exponential distributions by Brady. This model, however, was based on digitized voice statistics. Its applicability to packet voice with different audio content has not been examined before.

Packetization may change the traffic characteristics of packet voice, since both the techniques and implementations are different. In addition, while digitized voice in the traditional model was mainly concerned with telephone conversations, packet voice will be used for a wider range of new applications that may affect the traffic model, too. For instance, a significant amount of applications will be for information or entertainment distribution. The effect of audio contents other than conversations is not well understood, yet. It is, therefore, necessary to study the traffic characteristics of packet voice, and to examine the suitability of Brady's model to today's packet voice applications.

6.6.4.3 Audio Contents

In [48] spontaneous conversations between members or visitors of GTE Laboratories were evaluated. In addition to diversifying application environments and configurations, the topics of conversations were varied. Compared to a social conversation,

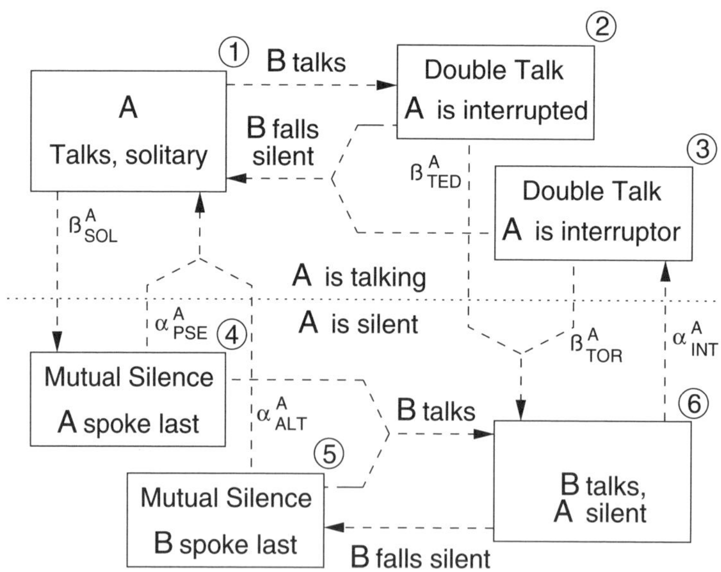

Figure 6.26: Six-state model

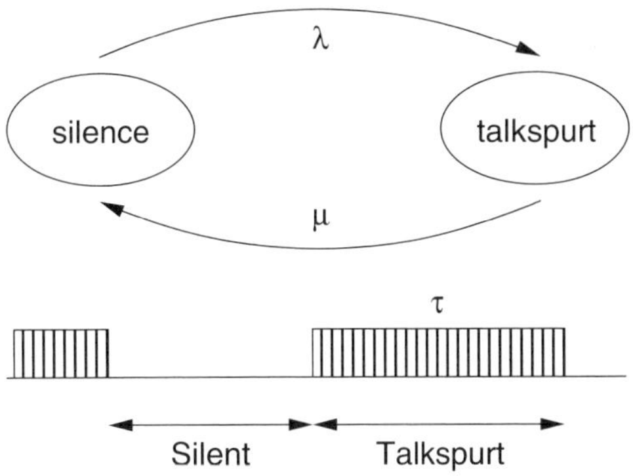

Figure 6.27: Single speaker model

Table 6.10: Traffic statistics of conversations [48]

Traces	**Talkspurt [s]** Mean	SDev	SDev/Mean	**Silence [s]** Mean	SDev	SDev/Mean	**Activity** Level
1	4.72	8.80	1.87	8.94	19.57	2.19	0.35
2	2.84	4.01	1.41	4.61	2.11	0.46	0.38
3	1.94	1.78	0.92	4.57	5.24	1.15	0.30
4	7.37	9.15	1.24	5.32	3.27	0.61	0.58
5	5.91	3.98	0.67	6.01	3.48	0.58	0.50
6	1.34	2.93	2.19	5.83	6.89	1.18	0.19
7	2.10	2.55	1.22	4.49	4.37	0.97	0.32
8	6.86	6.64	0.97	3.46	3.40	0.98	0.67
9	5.58	7.51	1.35	3.27	2.34	0.72	0.63
10	4.44	4.51	1.02	4.32	4.68	1.08	0.51
11	9.73	22.34	2.29	0.26	0.07	0.29	0.97
12	3.17	2.82	0.89	4.85	3.91	0.81	0.40
13	2.54	3.46	1.36	2.98	2.67	0.90	0.46
14	28.19	47.29	1.68	39.29	34.18	0.87	0.42
15	11.58	15.76	1.36	3.95	3.96	1.00	0.75
16	4.08	6.35	1.56	4.68	2.10	0.45	0.47
17	2.57	2.33	0.91	5.82	6.82	1.17	0.31
18	17.17	18.91	1.10	9.41	4.76	0.51	0.65
19	6.43	10.32	1.60	6.89	3.19	0.46	0.48
20	7.01	4.55	0.65	3.26	1.52	0.47	0.68
21	1.54	1.41	0.92	0.67	3.50	5.20	0.69
22	16.03	14.55	0.91	0.43	0.16	0.37	0.97
23	19.03	19.74	1.04	0.45	0.19	0.43	0.98
24	1.68	5.18	3.09	2.71	10.23	3.77	0.38

SDev = Standard Deviation

technical discussion may have more pauses, i.e., longer silences, caused by thinking or attending to other tasks such as retrieving a file, finding a reference and so on. It is also necessary to include the type of audio contents. The traditional traffic model was developed solely for telephone conversations. Packet voice, however, is serving

Table 6.11: Parameters for voice modeling with exponential distribution

	Single-Speaker model for digitized speech [41]	Traditional Packet Voice [48]	Packet Voice [48]
talkspurt	1.34 s	0.352 s	7.24 s
silence	1.67 s	0.65 s	5.69 s

a much wider range of applications. Prominent among them is information distribution. Examples may include delivery of news or entertainment programming, remote lecturing, retrieval of voice mail or e-mail in synthesized voice. They are referred to as *scripted speech*. Scripted speech may conceivably have different traffic characteristics from conversational applications. In Table 6.10 24 traces of Conversational Speech as specified by Deng can be found [48].

6.6.5 Packet Voice Traffic Model

The proposed traffic model for VoIP is based on the Single Speaker Model as described in Section 6.6.4.2 (see Figure 6.27).

In [48] the suitability of the six-state model for voice traffic was tested to verify the two null hypotheses: "Talkspurt period and silence period of a packet voice session are exponentially distributed". Deng's statistical analysis of the actual packet voice traffic indicates that only the average talkspurt and silence periods are much longer (see Table 6.11) than the values traditionally used, but the traditional model of exponential distribution is also statistically rejected in some cases. Traffic data reported in Deng's paper suggest that factors such as different speakers, topics and situations of a voice call may cause traffic patterns too diverse to be described by one single model expressed by the statement:

> The diversity of human behavior in oral communications between different individuals and under different situations suggests that there is no single statistical model that can fit all voice communications satisfactorily.

During talkspurt periods, the codec generates a CBR of 24 byte every 30 ms (based on the high data rate of the G.723, Section 6.6.2.1), or 20 byte at the low rate. Directly after the transition to the silent state a four-byte frame for the CNG is transmitted.

7
The (E)GPRS Simulator GPRSim

The (E)GPRS Simulator GPRSIM is a pure software solution based on the programming language C++. For implementation of the simulation model the *Communication Networks Class Library* (CNCL) [85] is used that is a predecessor to the *SDL Performance Evaluation Tool Class Library* (SPEETCL) [103, 31]. This allows an object-oriented structure of programs and is especially applicable for event-driven simulations. The complex protocols like LLC, RLC/MAC, the Internet traffic load generators and TCP/IP are specified formally with the *Specification and Description Language* (SDL) and are translated to C++ by means of the Code Generator SDL2CNCL [91] and are finally integrated into the simulator.

Unlike the usual approaches to building a simulator, where abstractions of functions and protocols are being implemented, the approach of the GPRSIM is based on the detailed implementation of the standardized protocols. This enables a realistic study of the behavior of EGPRS and GPRS. In fact, the real protocol stacks of (E)GPRS are used during system simulation and statistically analyzed under a well-defined traffic load.

The event control is performed by event handlers that are activated by arriving events and that send events to other event handlers after processing. The scheduling is done by a scheduler, which determines the order of processing of the events. Each event has a priority and a defined processing time. The simulation time advances in discrete steps, when all events, the processing time of which coincides with the current simulation time, are processed. Through this contemporaneity in the simulation process the simultaneous reaction of, e.g., several mobile stations to an event is represented realistically.

7.1 Structure of the (E)GPRS Simulator

The software architecture of the GPRSIM and the information flow between the modules is shown in Figure 7.1. The simulator comprises the modules MS, BS and SGSN, the transmission links, the load generator, session control modules, a graphical user interface for presentation and a module for statistical evaluation. Multiple instances of MS and BS can be generated and studied in a multicellular environment.

The formal stucture of MS, BS and SGSN is similar. They contain the implementations of the respective protocol stacks. The transmission links are represented by error models. While the G_b interface is regarded as ideal, block errors on the radio interface U_m can be simulated based on look-up tables, which map the actual *Carrier-to-Interference* (C/I) to a *Block Error Probability* (BLEP).

The generator comprises:

- Internet applications (WWW, e-mail, FTP, WAP, telematics, real-time and streaming applications)

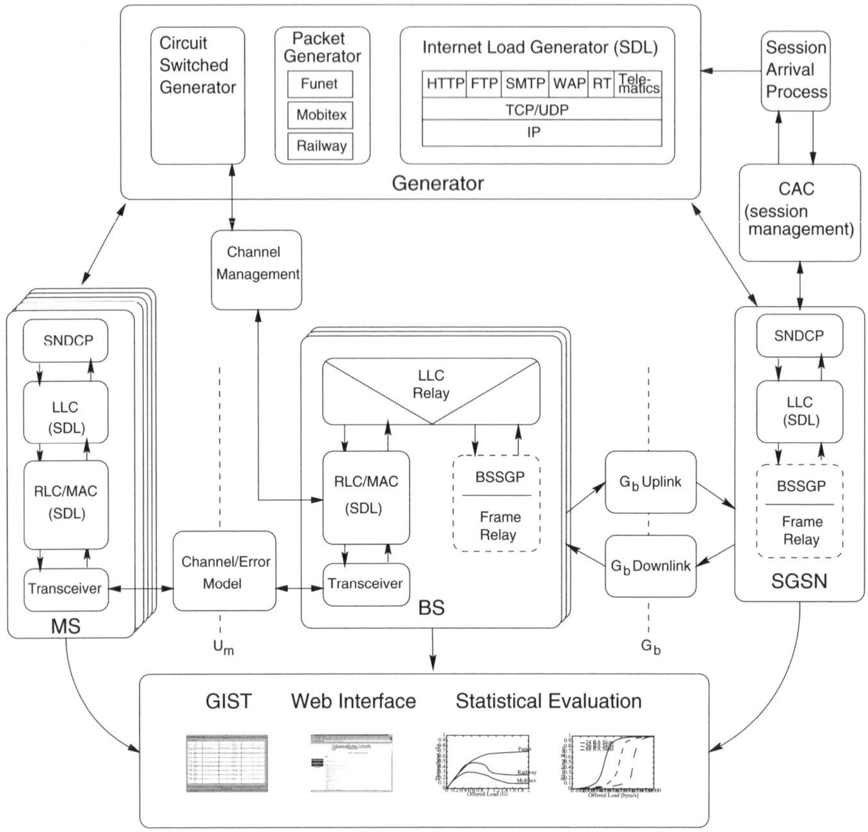

Figure 7.1: Structure of the (E)GPRS simulator

- the packet-traffic models *Finnish University and Research Network* (FUNET), Mobitex and Railway (see Section 2.2.1) proposed by ETSI for performance analysis
- a circuit-switched generator to model voice traffic coexisting with the packet oriented data traffic.

The module *Channel Management* supervises the physical GSM channels available in the respective cell and allocates channels for the GPRS resource management entity, if the channels are not used by circuit-switched services or if they are not allocated as dedicated PDCHs for (E)GPRS.

The layers BSSGP and *Frame Relay* (FR) are not represented in the GPRSIM because the focus was set on the radio interface. The respective classes do not provide any functionality and simply forward the service data units to the peer entity.

The output of the simulator comprises a graphical presentation of the protocol cycles and the statistical evaluation results of the performance measures.

In the following sections the different modules are presented with their functionality and interactions.

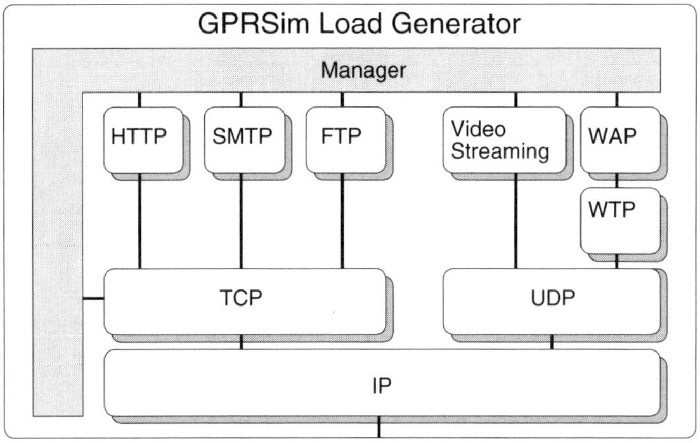

Figure 7.2: The GPRSIM load generator

7.2 Packet Traffic Generators

The GPRSIM comprises an Internet load generator based on the Internet traffic models for the applications WWW, FTP, e-mail, WAP, and Streaming (see Chapter 6). The transport and network protocols TCP, UDP, and IP are also implemented.

Besides the Internet load generator a packet generator based on the models FUNET (electronic mail), *Mobitex* (fleet management) and *Railway* (railway application) as proposed by ETSI [42] are implemented. These traffic models were proposed during the GPRS standardization process and have lost their importance for GPRS traffic modeling.

The conception of the GPRSIM *Internet Load Generator* is the specification of a system *Client* that behaves like a user running an Internet session, and a system *Server* representing the peer entity for the Internet service. For every MS one client and one server instance of the load generator are created. Below the application model the protocols TCP, UDP, WTP and IP are included. IP packets are generated and transmitted over the GPRS protocol stack.

7.2.1 Internal Structure

The internal structure of the GPRSIM Load Generator (see Figure 7.2) is derived from the ISO/OSI reference model. The *Application Layer* comprises the most important applications we know today like surfing in the *World Wide Web* (WWW) via HTTP, sending and receiving e-mails over SMTP/POP3 or downloading large files via FTP. Additional to these TCP-based applications there are UDP-based services.

The connection-oriented (TCP) and the connectionless transport protocols (UDP, WTP) belong to the *Transport Layer*.

Finally every TCP segment is encapsulated by the *Internet Protocol* (IP) belonging to the *Network Layer*.

Starting a new session the Manager chooses randomly between the applications based on configurable session probabilities. Then the client instance sends a request to the corresponding server to initiate a data transfer and the server confirms the request. The size and number of requests and answers depend on the chosen application type.

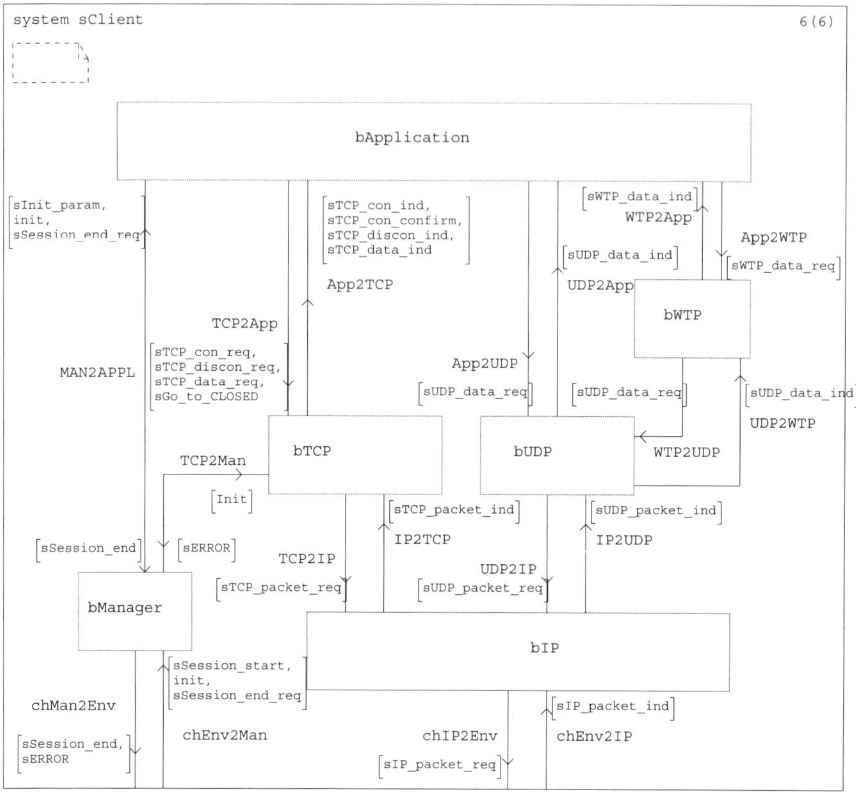

Figure 7.3: The block `sClient` in SDL

Furthermore the Manager is transmitting session control signals, e.g. `session_start` or `session_error`, from the different layers to the underlying `SDL_Process_environment`. Conversely the environment sends signals from the session control module, e.g., `init` or `session_end_request` to the corresponding entities.

7.2.2 Implementation

7.2.2.1 System Client

The system `sClient` represents the client for Internet traffic. Six different blocks are present inside the system (see Figure 7.3). While five of them determine the overall system behavior the sixth manages administrative purposes.

Inside the block `bApplication` one `process` for each traffic model exists (for details of the underlying traffic models see Chapter 6).

- At `sSession_start` the block `bManager` chooses randomly between the applications. The default probabilities of the different applications can be set via the configuration file.
- Receiving `sInit_param`, the block `bApplication` gets information about the protocol to be used. The specified SDL-process starts to initiate the application-related actions and requests objects from the Internet server instance according to the traffic model. Four parameters are generated by the `sClient`:
  - the number of Web pages or files to request,
  - the delay between each page/file,
  - the size of transmitted e-mails
  - the duration of a video stream.
- The blocks `bTCP` and `bUDP` respectively receive the generated segments and encapsulate them according to the protocol with the corresponding header. In case of TCP different phases like *slow start* or *congestion avoidance* are performed.
- After the block `bIP` has fragmented packets bigger than the Ethernet *Maximum Transfer Unit* (MTU) (1500 byte) it adds the IP header to datagrams and sends them to the environment.

Conversely, packets received from the environment via `sIP_packet_ind` are handed out to the block `bIP` and are passed through the protocol stack in the opposite direction.

7.2.2.2 System Server

The system `sServer` models the server of the Internet session. It is also composed of six different blocks and is similar in structure to the system `sClient`.

- Only the block `bApplication` differs from the client side, corresponding to the different functionality of load generation. The application-related processes have to generate random values for the size of requested objects, files, messages or e-mails.
- The process `pReal_Time_Server` inside `bApplication`, which is responsible for streaming video, is generating a whole sequence of RTP packets. The basis for these data is recorded traces of video sequences with different levels of motion, namely `Claire`, `Carphone` and `Foreman`.
- The different application protocols are distinguished by the protocol control received within the header as in the `sClient`.

7.3 Traffic Generator for Circuit-switched Services

The *Circuit-Switched* (CS) traffic generator generates events with an interarrival time determined by a negative-exponential distribution. These events correspond to calls initiated in a cell. When one event arrives, directly the next event is generated to realize the arrival process behavior. A channel assign request is sent to the channel

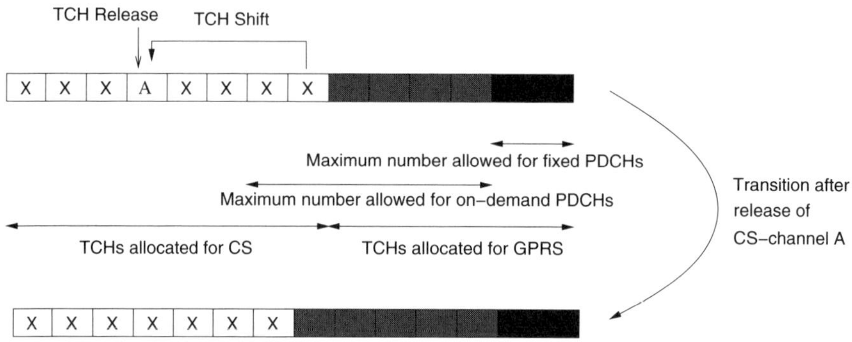

Figure 7.4: TCH and PDCH allocation table

management module. If there is a traffic channel available, the request is confirmed and the call duration, also negative-exponentially distributed, is determined. After the call duration an event is sent to the channel management module to release the assigned channel. The traffic value in Erlang is given by the two configurable mean values of the call interarrival and the call duration times.

7.4 Channel Management

The Channel Management module in the GPRSim applies dynamic channel allocation to control the pool of GSM traffic channels (TCHs) for GPRS and GSM applications as described in [17]. A TCH can be used for either a circuit-switched speech connection or a PDCH. Channel Management communicates with the CS traffic generator and the GSM/GPRS Radio Resource Management (RRM) instance that is located in the RLC/MAC layer. The state of this module is defined by the number of TCHs available to a radio cell, the maximum allowed number of fixed PDCHs, the maximum allowed number of on-demand PDCHs and the actual number of CS connections. Based on these parameters Channel Management determines the number of PDCHs currently available for GPRS. CS connections are prioritized with preemption, i.e., a new CS request interrupts and uses a PDCH used so far by GPRS immediately, if no other TCH is free in the cell. All TCHs that are not used by CS connections are available for GPRS, if the number of PDCHs does not exceed the sum of the maximum allowed number of fixed and on-demand PDCHs. It is assumed in the simulations presented that TCHs used for CS connections are placed adjacently beginning with the TDMA channels of carriers not used for GPRS. If a CS connection (say A) is released, the TCHs are shifted in a way so that TCHs and PDCHs are adjacent again after release; see Figure 7.4. In a real system this goal could be reached by intracell handoff of only one TCH. This mechanism is known as repacking. A transition, i.e., when the number of PDCHs available for GPRS changes, is immediately indicated to the RRM of RLC/MAC.

7.5 Quality of Service Management

In this section, a description of the concepts developed to model the *Quality of Service* (QoS) management functions presented in Section 4.3 and Section 4.4 is provided, as well as a brief description of the relevant implementation [107].

7.5.1 QoS Profile and Aggregate BSS QoS Profile

A data structure is needed for the data defining the various aspects of QoS within the GPRS system regarded.

The class `QoSProfile` contains all parameters necessary to specify a mobile user's subscribed QoS, a mapping function of the application type onto an appropriate traffic class (see Section 4.4.5), and the traffic classes' definition in terms of the following service classes (see Section 4.4.3):

- precedence class
- reliability class
- delay class
- mean throughput class
- peak throughput class

The service classes correspond to the relevant values representing the values listed in Section 4.4.3.

`QoSProfile` offers an interface to initialize these values according to the simulation scenario requirements, to modify them during PDP context activation or modification procedures, and to set appropriate service class values for application traffic flows.

However, the service class values are not utilized for scheduling purposes—this is done solely on behalf of ABQPs, as described in Section 3.2. The ABQP, thus, is considered a single parameter with multiple data transfer attributes that is exchanged between SGSN and BSS during PFC-context related procedures. For implementation purposes, it is not necessary to combine subscription and traffic class—which will define the ABQP—into one parameter. They are instead encapsulated in a C++ `struct`, as shown in the following source code.

```
typedef struct aggregateBSSQoSProfileType : public CNObject
{
  int              tlli ;
  subscriptionType subscription;
  trafficClassType  trafficClass ;
};
```

For QoS differentiation both application classes (Conversational, Streaming, Interactive and Background) and subscriber classes (Premium, Standard and Best-Effort) are considered.

7.5.2 PDP Context

Since communication paths between the GPRSim modules involved in PDP-context related signaling are modeled only logically (see Section 7.5.4), there are few fields of the PDP context *Information Element* (IE) necessary to be modeled within the scope of the simulator. Hence, the class `PDPContext` simply contains the QoS Sub, QoS Req, QoS Neg fields, and the PDP address.

```
class PDPContext : public CNObject
{
  public:     //***** Constructors *****/
    PDPContext();
    PDPContext(int);
    PDPContext(const PDPContext&);
    PDPContext(CNParam *) {}

  public:     //***** Public interface *****/
    int           PDPAddress;

    QoSProfile* QoSSub;
    QoSProfile* QoSReq;
    QoSProfile* QoSNeg;

    PDPContext& operator=(const PDPContext&);

  private:    //***** Internal private members *****/

// {{{ Member functions required by CNCL...
};
```

7.5.3 GPRS Register (GR)

The class `GR`, representing the *GPRS Register* (GR) within the HLR, is responsible for the storage of subscriber related QoS profile data. It provides an array of pointers to objects of type `QoSProfile`. This array's size is equivalent to the number of MSs for the respective simulation scenario.

The tasks of `GR` comprise the initialization of the array of QoS profiles according to the relation of *premium* to *standard* to *best-effort* subscribers specified for the respective scenario, and the delivery of subscriber QoS profile data on request from an SGSN handling a PDP context activation. Moreover, the class `GR` has to provide an interface to the generator to allow traffic load generation in conformance with the subscriber QoS profiles valid for the simulation scenario under examination.

7.5.4 Connection Admission Control (CAC)

As outlined in Section 4.4 the *Connection Admission Control* (CAC) function plays a decisive role in QoS-oriented traffic management in a GPRS environment. Within the scope of the GPRSIM, it has to fulfill various tasks, from *Admission Control* (AC) and QoS profile negotiation to radio resource status monitoring. Within the context of the GPRSIM the class CAC represents the QoS-management related functions of both the SGSN and the GGSN. Communication paths from MS via BSS to the CAC module within the GSNs are recreated only logically and do not utilize any physical signaling resources of the system.

As already shown in Figure 4.29, the module CAC is responsible for handling PDP context activation and deactivation requests, keeping track of traffic load and radio resource utilization status within the radio cell on behalf of BSSGP flow control information from the BSS, and performing PDP context renegotiation. Thus, there are interfaces necessary for communication with the modules GR, generator, BS transceiver, and the RLC/MAC modules via the SDL process environment.

The message flow between the module CAC and the other GPRSIM modules involved in the AC process is summarized in Figure 7.5.

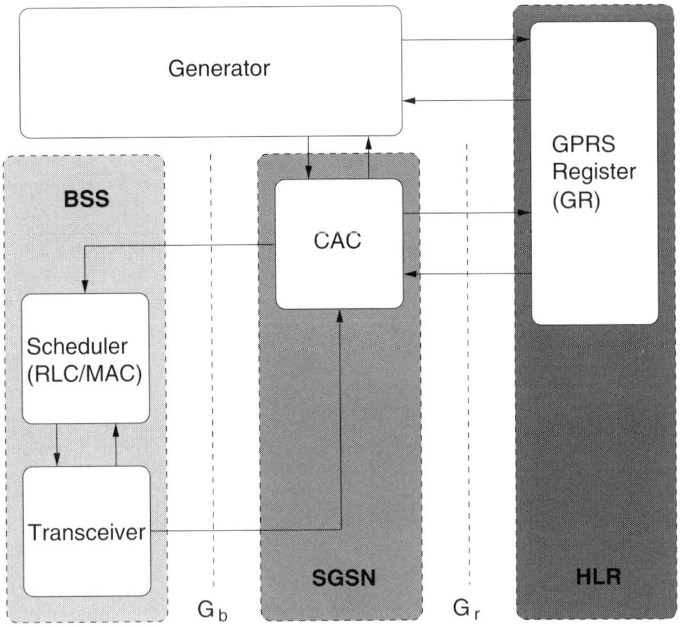

Figure 7.5: Traffic flow between the GPRSIM modules involved in QoS management

The module CAC must provide methods and objects to store, update, and evaluate radio resource utilization data as well as traffic-flow-related QoS profile data (*Traffic Flow Template*, TFT). Further, it has to perform *Admission Control* on the basis of this data, and react to changing resource utilization in a radio cell by engaging PDP context renegotiation procedures.

To perform *Admission Control*, there is an AC policy to be implemented in a way that all active traffic flows can be served according to the QoS profiles negotiated. Preferably, there is only a limited number of privileged connections allowed simultaneously if the QoS guarantees given by subscription do not run the risk of being violated.

Furthermore, any *Standard* traffic should receive the resources necessary to meet its QoS requirements. Thus, it might be preferable to reject or downgrade a PDP context activation request rather than to endanger the quality of all sessions. On the other hand, it might be advantageous to pre-empt background or even interactive traffic flows to allow for an additional *Conversational* traffic flow to be admitted.

Therefore, PDP context activation requests are differentiated according to their subscribed QoS profiles and their traffic classes. The CAC function first examines whether there are additional PDP contexts allowed for the requested QoS profile. If this proves to be true, in a second step it is decided on the basis of the radio cell resource utilization whether to grant access or not. For this purpose, it is necessary for the CAC to keep track of the radio cell resource utilization status. The BSS channel management module thus has to regularly transmit according data.

To avoid a total withdrawal of resources from the *Standard* traffic classes with lower QoS requirements, e.g., other than *Conversational*, there is a share reserved for this kind of traffic from the pool of radio resources in the cell. In general, all resources

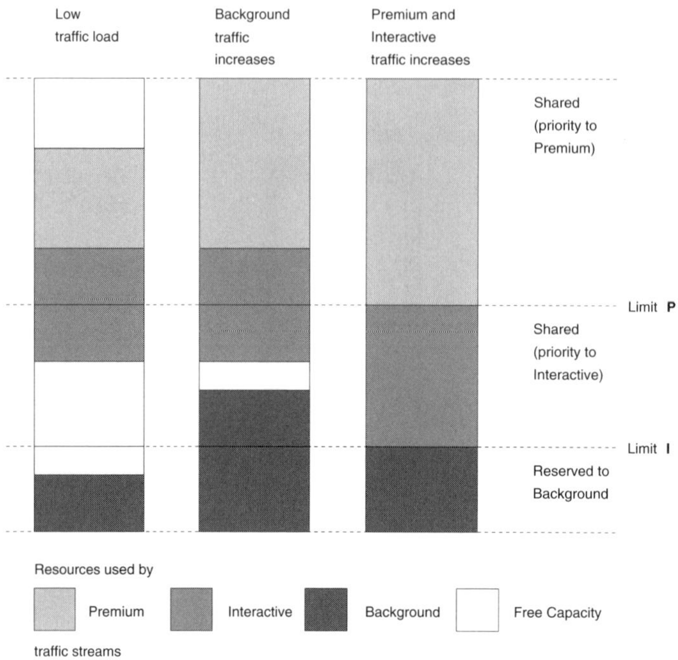

Figure 7.6: Admission control policy (example)

are open to traffic of any kind. In times of high load, however, traffic flows with more demanding QoS requirements are allowed to pre-empt flows belonging to applications with lower QoS requirements, but only up to a certain limit (see Figure 7.6), where P and I represent the appropriate limits for the number of accepted flows per class.

If this limit is reached, the requested QoS is not granted, but rather degraded to the next-lower-prioritized class. Within the framework of this example, no deactivation of the PDP context by an MS, because of not receiving the QoS requested, is initiated.

7.5.5 Scheduler

On behalf of the QoS profile negotiated the BS MAC layer has to perform scheduling of the radio blocks. For evaluation purposes, only the downlink direction is regarded within this framework, since the traffic models used within the generator module of the GPRSIM produce a considerably higher amount of data traffic for this direction, compared with the uplink. Thus, the downlink direction may be regarded as the bottleneck in respect of resource utilization. Traffic-class based uplink scheduling is performed during USF assignment and channel request scheduling.

For future implementation of further traffic generator modules, e.g., for *Conversational* applications like VoIP, it is necessary to also regard the uplink QoS.

The scheduling mechanism implemented for both the downlink and uplink directions follows a three-stage principle (see Figure 7.7). First, incoming radio blocks are classified into one of three queues according to the QoS subscription associated with the respective traffic flow. It is differentiated between *Premium* (*Gold Card*) service, *standard* service and *best-effort* service.

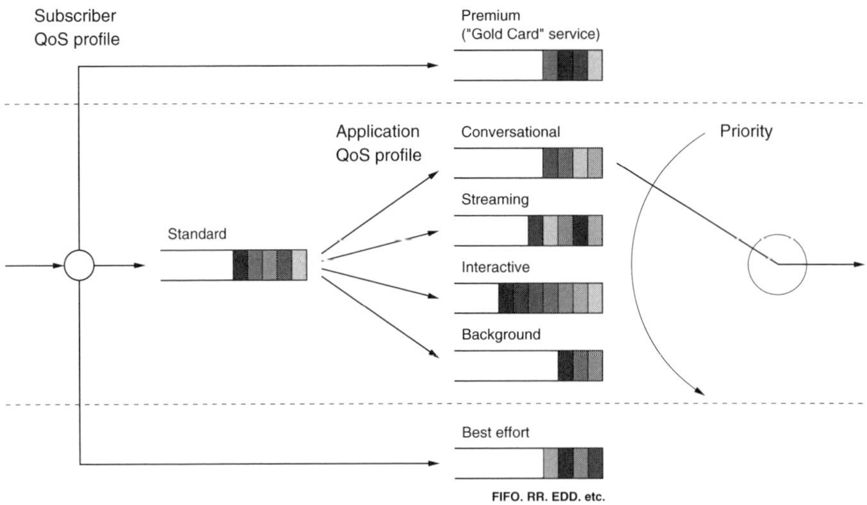

Figure 7.7: Principle of the scheduling function located in the BS MAC layer

The second stage is only valid for *standard* service traffic. Regarding a packet's application QoS profile, the appropriate traffic class queue is chosen from *Conversational*, *Streaming*, *Interactive*, or *Background*. Best-effort traffic from the first stage is put into a fifth queue. Within the traffic class queues, packets are scheduled according to their TLLI and a *Round Robin* (RR) algorithm. Other algorithms than RR may also be implemented on this stage, e.g., *First Come First Served* (FCFS) or *Earliest Due Date* (EDD).

The third stage is built by a simple priority mechanism, serving the traffic class queues in order from highest priority *(premium)* to lowest priority *(best-effort)*. To allocate bandwidth to the traffic classes *Weighted Round Robin* (WRR) or *Weighted Fair Queueing* (WFQ) schemes can be applied in order to avoid blocking of the lower prioritized classes.

All three stages are part of the GPRSim RLC/MAC layer module and specified in SDL.

7.6 The SNDCP Layer

The *Sub-Network Dependent Convergence Protocol* (SNDCP) class has two main functions. First TCP/IP header compression is performed. Then the IP packets coming from the IP layer are adapted to the LLC layer. As the *Maximum Transfer Unit* (MTU) in the LLC layer is 1520 byte, bigger IP packets have to be segmented and reassembled by SNDCP. In addition SNDCP has a queueing function in the GPRSim. With the function `put_ip_packet` the given parameter `ip_packet` is segmented into SNDCP PDUs and queued as LLC *Interface Data Unit* (IDU). When the LLC layer is ready to send new data, an LLC IDU can be dequeued.

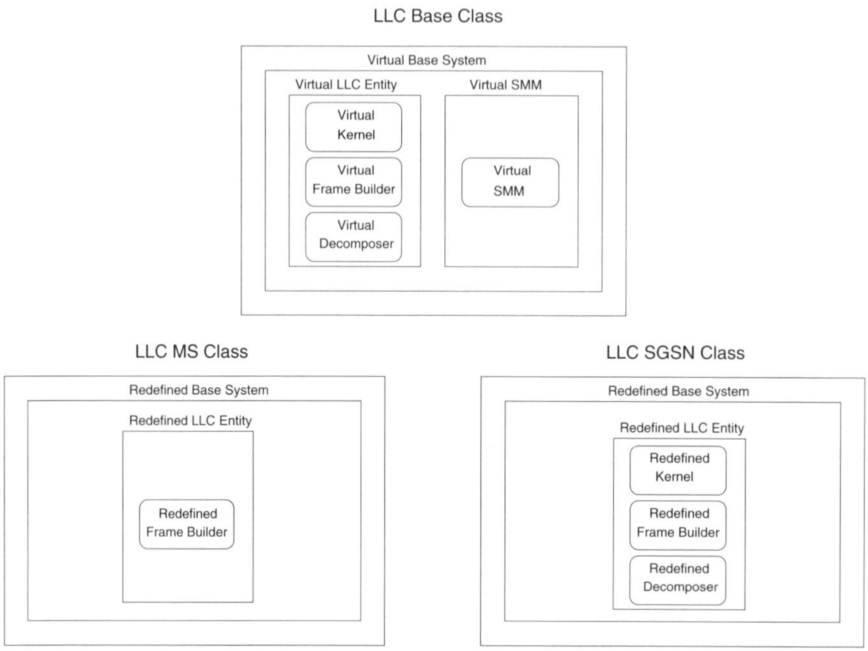

Figure 7.8: Structure of the LLC layer as an SDL92 system

7.7 The LLC Layer

The LLC protocol is also specified in SDL (see Figure 7.8). Since the simulator is realized in C++ with the utilization of the CNCL, the SDL specification is translated to C++ code with the code generator SDL2CNCL [91]. The generated code has been integrated into the GPRSIM.

The class `SDLProcess_SystemManagement` represents the interface with the network layer. It prepares the signals received from the SDL process environment for the LLC layer and sends the associated signals to the environment.

The classes `SDLSystem_LLC_MS` and `SDLSystem_LLC_SGSN` represent the LLC entities of the MS and the SGSN. They contain several processes that organize and control the transfer procedures. The class `SDLProcess_Kernel` is the kernel of the LLC layer. Here received frames and service primitives are processed and the related reactions are performed. In the kernel it is decided which frames can be sent and which services are offered to the network layer. In the course of a transfer the kernel enters different well-defined states, which determine the reaction to received signals. The states are presented in Figure 7.9. To guarantee a clear presentation, sender and receiver are presented separately. Transitions between states are caused by specific events that are pictured by arrows.

The kernel informs the class `SDLProcess_FrameBuilder` which frames have to be sent. This class composes the frames according to the frame type or command and transfers them with an `RLC/MAC`- or `BSSGP-DATA-REQ` to the underlying protocol instances.

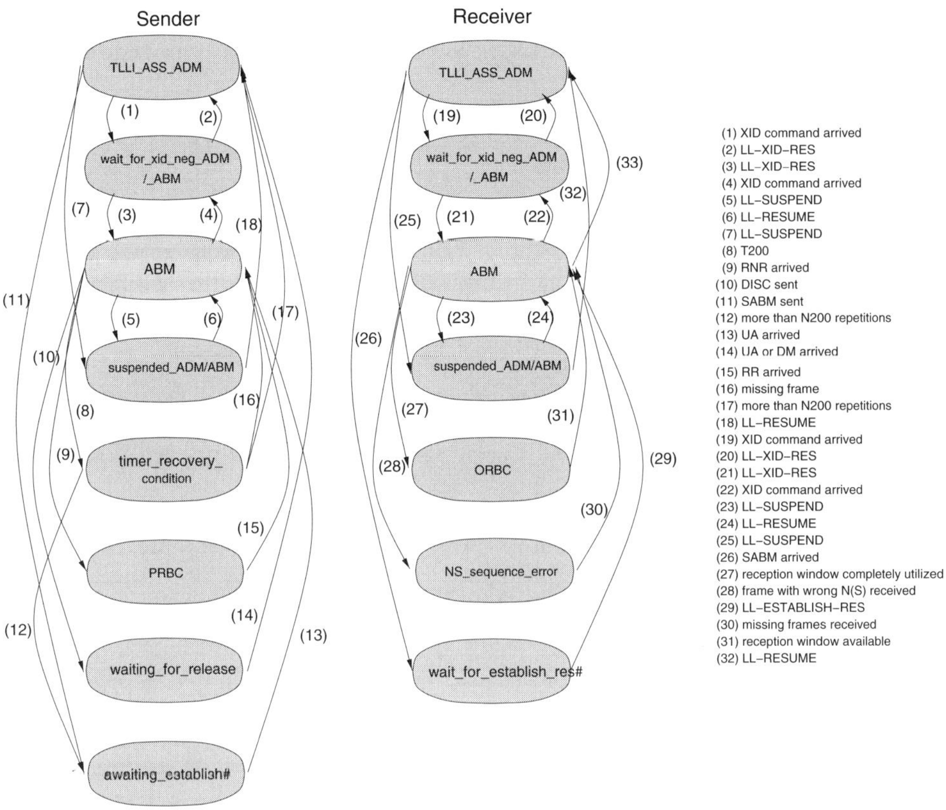

Figure 7.9: State machine of the process LLC entity

The class SDLProcess_Decomposer receives an LLC PDU by an RLC/MAC- or BSSGP-DATA-IND. After evaluation of the control field the data of the information field is forwarded to the LLC kernel.

In the following paragraph the states and events of Figure 7.9 are described.

TLLI_ASS_ADM: In this state the LLC entity waits for the reception of a frame or an LLC service primitive. The transfer of *Unnumbered Information* (UI) frames or the exchange of Exchange Identifier (XID) parameters is not possible in this state.

awaiting_establish: An LL-ESTABLISH-REQ primitive from the network layer has arrived. The LLC layer has sent a Set Asynchronous Balanced Mode (SABM) message and waits for the acknowledgement (Unnumbered Acknowledge (UA)). After that it moves into the Asynchronous Balanced Mode (ABM) state.

wait_for_establish_res: A SABM was received and the network layer was informed by an LL-ESTABLISH-IND. The LLC entity waits for an LL-ESTABLISH-RES from the network layer. After reception a UA response is sent and the new state is ABM.

ABM: In this state the acknowledged data transfer is possible. The LLC entity waits for an `LL-DATA-REQ` from the network layer or the reception of an `Information` (I)-frame from the peer entity.

wait_for_release: The LLC entity has received an `LL-RELEASE-REQ` from the network layer. After sending the `Disconnect` (`DISC`) message it waits for a `UA` or a `Disconnect Mode` (`DM`) response from the peer entity.

wait_for_xid_neg_ADM: After receipt of an `XID` command in `Asynchronous Disconnected Mode` (`ADM`) and the forwarding to the network layer the entity waits for the negotiated parameters from the network layer. After that the parameters are transmitted to the peer entity.

wait_for_xid_neg_ABM: After receipt of an `XID` command in `ABM` state and the forwarding to the network layer the entity waits for the negotiated parameters from the network layer. After that the parameters are transmitted to the peer entity.

suspended_ABM: The LLC entity is temporarily suspended and waits for an `LL-RESUME-REQ` from the network layer. After that it goes to `ABM` state.

suspended_ADM: The LLC entity is temporarily suspended and waits for an `LL-RESUME-REQ` from the network layer. After that it goes to `ADM state`.

PRBC: In `Peer_Receiver_Busy_Condition`, the peer entity is unable to receive further I-frames.

ORBC: In `Own_Receiver_Busy_Condition`, the peer entity is unable to receive further I-frames, but can transmit I-frames.

timer_recovery_condition: Timer `T200` has expired and the LLC entity calls for certain frames or commands from the peer entity.

NS_sequence_error: An I-frame with a wrong sequence number was received. The LLC entity calls for the missing frames and goes back to `ABM` after reception.

7.8 The RLC/MAC Layer

This protocol implementation models the RLC/MAC Standard GSM Rec. 04.60, Release 99, Version 8.2.0 [16]. The full functionality of RLC and MAC is represented. The architecture of this layer in MS and BS is similar, but the functionality of the processes is different. In the following the structure and the functions of the BS MAC layer are described. After that the differences in the MS MAC implementation are outlined.

7.8.1 RLC/MAC in the Base Station

In Figure 7.10 the structure of the RLC/MAC layer in the base station and the interaction between the functional processes and the environment are shown.

After the start of a simulation run the process `prBS_Config` obtains the configuration parameters from the environment and creates and initializes the instances `prBS_RLC_entity`, `prBS_DL_FSM` and `prBS_UL_FSM` for every mobile station associated with this base station. The other processes exist only once in the system. The

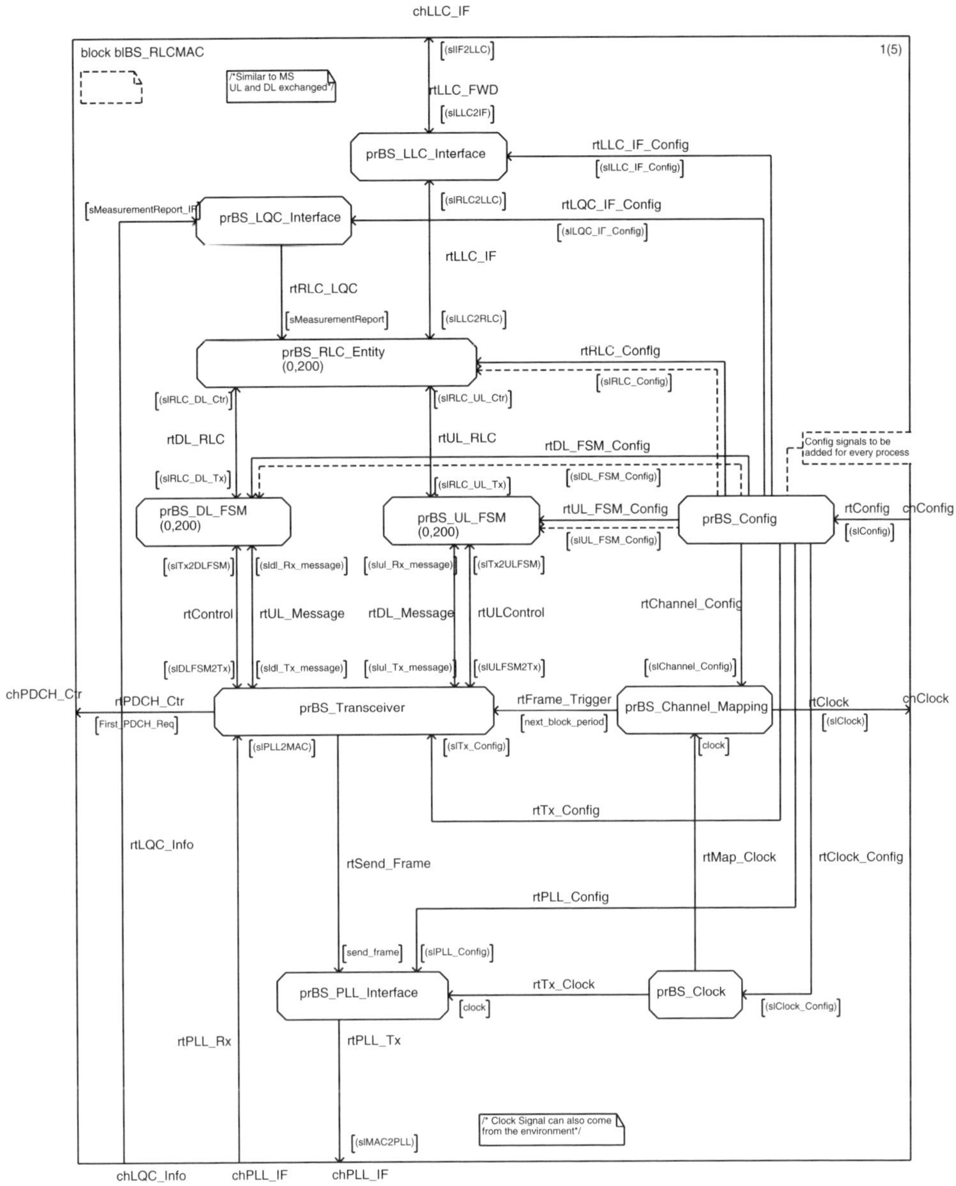

Figure 7.10: RLC/MAC base station architecture

data flow in the system is defined as follows. In the transmit direction LLC PDUs are requested from the higher layer always if the LLC queue is not empty and if there is enough space for new data to transmit in the send buffer of the RLC entity.

If necessary a new downlink TBF initiation is requested to the downlink *Finite State Machine* (FSM). The control blocks, which are part of the TBF establishment process, are given to the `prBS_Transceiver` and are transmitted as soon as radio resources are available. As soon as a TBF is established RLC blocks can be transmitted if downlink resources are available. These resources are managed by the BS transceiver, which requests new RLC blocks by asking the related RLC entity for new blocks to transmit. The order of RLC entities that are requested is defined by the scheduling function.

Complete coded radio blocks for each block period are given to the `prBS_PLL_Interface`. This process forwards one block after the other to the environment at the correct time regarding the multiframe structure. In the receive direction radio blocks are received by the BS transceiver from the environment and are forwarded to the related uplink or downlink FSM. If these radio blocks contain RLC blocks, i.e., data or acknowledgement blocks, they are forwarded by an FSM to the RLC entity. RLC data blocks are saved in the receive window and reassembled into LLC PDUs, which are finally given to the LLC layer.

In the following the internal functionality of the processes is described.

7.8.1.1 prBS_Config

The Configuration process gets protocol settings and scenario parameters from the environment. Processing these values the process creates the requested number of process instances for each associated MS in the cell. After that all processes are initialized with configuration parameters. Through this mechanism the processes can be configured from outside the system.

7.8.1.2 prBS_LLC_Interface

The LLC interface ensures the communication between the RLC entity of the logical link and the related LLC queue. In the transmit direction signals from an LLC instance are mapped onto the RLC entity associated to this logical link. In the receive direction reassembled LLC PDUs as well as other interworking signals from the RLC entity are forwarded to the associated LLC entity or queue.

7.8.1.3 prBS_LQC_Interface

The LQC interface forwards measurement reports about the link quality to the related RLC entity.

7.8.1.4 prBS_RLC_Entity

The RLC entity provides the transmission service for LLC PDU between MS and BS. An RLC connection involves two peer entities. It uses the radio resources offered by the MAC function. A selective repeat ARQ protocol is used to ensure high reliability. Each RLC endpoint has a receive window to receive the RLC data blocks and a transmit window to transmit the RLC data blocks. For GPRS the window size is 64. This means that a maximum of 64 data blocks can be transmitted until the oldest block in the transmit window must be acknowledged to allow transmission of further blocks. For EGPRS the WS depends on the number of allocated PDCHs. The value ranges from 64 to 1024.

Measurement reports in the form of a proposed optimal *Modulation and Coding Scheme* (MCS) in EGPRS, or *Coding Scheme* (CS) in GPRS, are received every 60 ms. The respective state machine is shown in Figure 7.11. After initialization the process is in state `Start`.

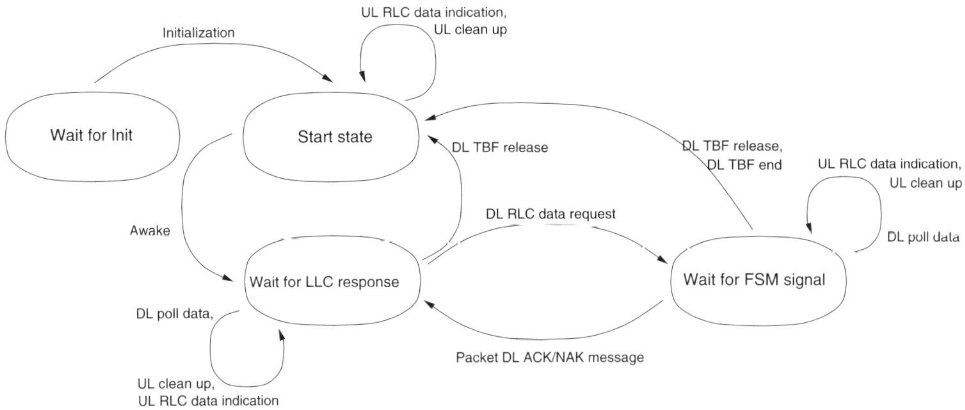

Figure 7.11: State machine of the process RLC entity at the base station

Transmit Procedures The LLC layer informs the RLC entity about the aim of sending LLC PDUs. After the TBF establishment of the MAC FSM, the RLC entity asks the LLC layer for a number of data octets whose value depends on the available free places in the send buffer and the used channel coding scheme. The CS used for each transmission is either a fixed default value or chosen based on the current channel quality (LA). After the requisition of data, the RLC entity waits for LLC response.

The LLC entity delivers a maximum number of PDUs according to the requested octets that are segmented into RLC data units. If the contents of an LLC PDU do not fill an integer number of RLC data units, the beginning of the next LLC PDU is placed within the final RLC data unit of the actual LLC PDU. No padding is necessary unless the last LLC PDU in the TBF does not fill an integer number of RLC data units.

The RLC header is generated and the RLC data blocks are stored in a send buffer, where it waits for transmission. For GPRS, the send buffer can contain a maximum number of 128 (*Sequence Number Space* (SNS)) RLC data blocks. For EGPRS the SNS range is 2048. Now the RLC entity waits for FSM signals.

When resources for data transmission are available, the MAC function polls for data and is served directly with blocks from the send buffer. The moment an RLC data block is given to the MAC the corresponding element in the acknowledge state array `V(B)` is set to `PENDING_ACK`. The elements in `V(B)` indicate the acknowledgement status (`NACK`, `PENDING_ACK`, `ACK`, `INVALID`) of the related RLC data blocks.

The data blocks are taken out of the send buffer in the following order: First, RLC data blocks whose corresponding elements in `V(B)` have the value `NACK` are delivered to the MAC, starting with the oldest one. If no data block with a `NACK` element exists and the transmit window is not stalled, the RLC data block with `BSN = V(S)` is selected. The send state variable `V(S)` denotes the block sequence number of the next in-sequence RLC data block to be transmitted. If the transmit window is stalled or there is no new block to send, the oldest RLC data blocks whose value in `V(B)` is `PENDING_ACK` are sent again.

For each RLC data block sent a timer is started. Only when it is expired is a negative acknowledgement for the corresponding RLC data block accepted. The reason is that directly after an RLC data block is retransmitted (e.g., `PENDING_ACK` - for reason see above) a Packet ACK/NACK message with a negative acknowledgement of the previous transmission can arrive. Then it is not reasonable to set the corresponding element in `V(B)` to `NACK` because this results in a direct retransmission of the same RLC data block.

The transfer of Packet ACK/NACK messages is managed by the RLC entity of the base station. In the downlink direction an acknowledgement message is demanded by the poll bit in the RLC header of the downlink data blocks. An ACK/NACK message is sent in the following cases:

- The receiver has received a configurable number of radio blocks.
- The *Stall Indicator* (SI) in the received data block indicates that the transmit window is stalled.
- All blocks within the TBF are received and the next ACK/NACK message is the last within the TBF.

After the receipt of a `PaDowACK/NACK` message the values of `V(B)` are updated from the values received from the peer entity in the RBB of the acknowledgement message. The transmit window is moved and the LLC layer is asked for new data octets. If the LLC layer has no more LLC PDU to transmit and all previous PDUs are successfully transmitted, the RLC entity initiates a regular TBF end and changes into the `Start` state.

An abnormal release during an active TBF can be performed by the MAC or by the RLC. The RLC entity initiates a TBF release, if no `PaDowACK/NACK` message has arrived in a configurable time after it has sent the last RLC data block in a downlink TBF.

An abnormal release which is performed by the MAC is indicated to the RLC. The transceiver is initiated and goes into the `Start` state where it has to wait for a new request from the LLC layer. If the LLC layer has no more LLC PDU to transmit and all previous PDUs are successfully transmitted, the RLC entity initiates a regular TBF end and changes into the `Start` state.

The GPRSIM operates with open-ended TBFs. The last RLC data block in a downlink TBF is indicated by setting the FBI. The logical connections between the RLC entities are different from the TBFs provided by the MAC. Delayed termination of the TBF context is provided. Before delayed termination is done the TBF context of the MAC layer can be used to transmit data after a short inter-packet-train gap.

Receiver Procedures In the `Start` state uplink data can be received at any time. RLC data blocks are collected in the receive buffer until all RLC data blocks comprising an LLC PDU have been received. As soon as an LLC frame is successfully transmitted across the RLC layer, it is reassembled and forwarded to the LLC layer. For each RLC data block which is received, the corresponding element in the receive state array `V(N)` is set to `RECEIVED`. The elements in `V(N)` indicate the receive status (`RECEIVED`, `INVALID`) of the related RLC data blocks.

When an acknowledgement message has to be sent, a Packet ACK/NACK message is composed. It contains a *Starting Sequence Number* (SSN) and a *Received Block Bitmap* (RBB) according to the standard.

In the case of an abnormal release of the uplink TBF, the receive buffer is initialized and the receiver is ready for processing a new uplink TBF.

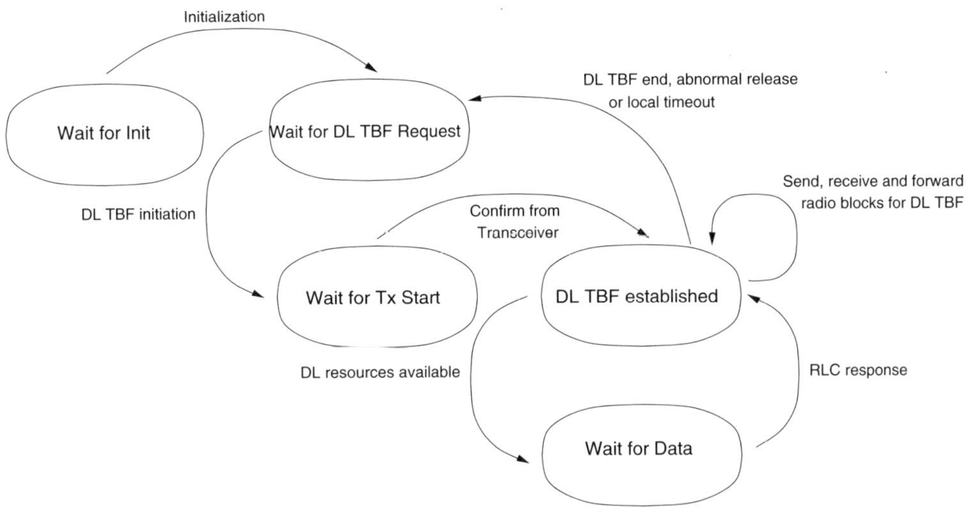

Figure 7.12: State machine of the process downlink FSM at the base station

7.8.1.5 prBS_DL_FSM

The downlink FSM manages the state sequence and behavior of the downlink TBF. It realizes functions to receive RLC/MAC control and data blocks. The state machine is shown in Figure 7.12.

After initialization the process waits for a downlink TBF request by the RLC entity. This corresponds to an idle state where no TBF exists. When an initiation arrives, a `PaDowAss` message is coded and sent to the transceiver. After that the process waits for the confirmation of the transceiver process implying that the downlink TBF setup has been performed. The process goes into the state `DL TBF established` after this confirmation. In this state the transmission and reception of downlink associated RLC/MAC blocks are possible. If the transceiver has allocated radio resources for a logical link the related downlink FSM is notified. On this polling signal an RLC data block from the related RLC entity is demanded and after the RLC response the data block is forwarded to the transceiver process or the transceiver is notified that there is no data to transmit. For every state it is ensured that a timer resets the process into the `idle` state, if the process stays a certain time in a wait state without receipt of a response signal. If a regular TBF end is indicated by the RLC entity, the TBF context is released after a delay time in order to leave the TBF available to new LLC data which has to be transmitted after a short inter-packet-train gap.

7.8.1.6 prBS_UL_FSM

Similar to the process downlink FSM the uplink FSM manages the state sequence and behavior of the uplink TBF. It manages the receipt of RLC/MAC control and data blocks. The state machine is shown in Figure 7.13.

After initialization the process moves into the `idle` state and waits for an uplink TBF request in form of a `PaChaReq` message sent by the peer entity and received and forwarded by the transceiver process. After decoding this control block the process continues with the procedure of one-phase or two-phase access. Performing one-phase

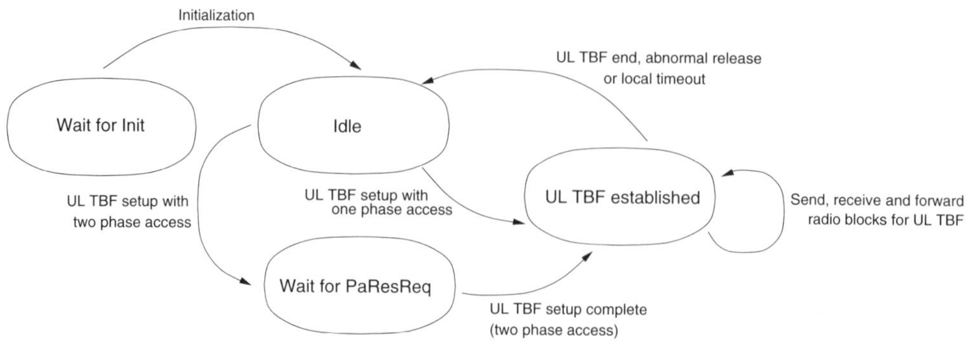

Figure 7.13: State machine of the process uplink FSM at the base station

access the process answers with a `PaUplAss` message, which is coded and given to the transceiver. After that the uplink TBF is established. Performing two-phase access first a `PaUplAss` is coded and given to the transceiver and the process waits for a `PaResReq` message from the peer entity. After reception of this message the process sends a `PaResAss` message to the transceiver. Then the uplink TBF is established. In this state RLC/MAC blocks can be sent, received and forwarded to the upper and lower processes. As in the downlink FSM errors are resolved by resetting the process into the idle state if a timeout occurs. If a regular TBF end is indicated by the RLC entity, the TBF context is released after a delay time in order to let the TBF open for new LLC data which has to be transmitted after a short inter-packet-train gap.

7.8.1.7 prBS_Channel_Mapping

This process recreates the multiframe structure of (E)GPRS. It receives the clock signal which corresponds to the beginning of each time slot. For each time slot this process decides which logical channel is represented by the current time slot for the uplink and the downlink. This is easily performed by a channel map which maps a position in the multiframe structure on a logical channel identifier. The channel mapping is performed according to the description in Section 2.2.5.3. The optional parameters such as the number of PCCCHs and PACCHs within the PDCHs are configurable and are defined during the initialization phase.

Additionally the process sends a trigger signal to the transceiver in order to indicate a new block period and indicate the channel parameters for this block period, such as the number of PDCHs available and the type of logical channel of each block in the block period. Other trigger signals indicate to the environment whenever a new slot, a new TDMA frame and a new 52 multiframe begins.

7.8.1.8 prBS_Transceiver

The principal data structures in the transceiver are

- A resource database, which stores the current information about the radio resources available and allocated (e.g., TFI, USF).
- A TBF information table, which stores the states and parameters of the uplink and downlink TBFs of each logical link.

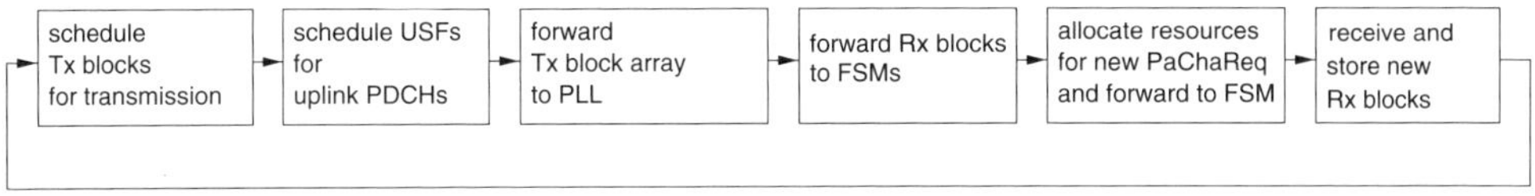

Figure 7.14: Task sequence in a block period at the base station

- Message queues for common and associated control messages to transmit.
- Buffers for received control blocks, which have to be processed and forwarded.
- Queues containing `PaChaReq` messages, which wait for resource allocation.
- TBF history queues, which contain scheduling information.

As part of the process the transceiver receives radio blocks (Rx blocks) and processes the header information of these blocks. It extracts the logical link that is addressed, updates the data structures depending on the message type and its parameters and stores the blocks in the receive buffer. Received blocks are forwarded later to the uplink or downlink FSM of the related logical link. If a `PaChaReq` message arrives, resources have to be allocated to an uplink TBF regarding the multislot class of the MS. If no resources are available, a queuing notification is stored in the common control queue for transmission.

Control messages requested for transmission from the uplink or downlink FSMs (Tx blocks) are stored in the common or associated control queue. If a `PaDowAss` message is requested for transmission by the uplink FSM for a downlink TBF establishment, downlink resources have to be allocated regarding the multislot class of the MS and coded in the `PaDowAss` message, which is stored in the common control queue.

Always when a new block period is indicated by the channel mapping process, a number of radio blocks for the downlink are scheduled depending on the number of available PDCHs. For the uplink an uplink TBF is scheduled for each uplink PDCH available. The related USF field of each downlink radio block is set to the USF of the scheduled uplink TBF. The scheduling function works as follows. For each PDCH that is available in the current block period the channel type is examined. According to this channel type common control or associated control messages are taken from the message queues. If no control messages are scheduled, data blocks are demanded over the downlink FSM from the RLC entity. The sequence of the logical links that are polled for data blocks to transmit is determined by a round robin and priority scheduling strategy. The USF for each radio block is determined by another scheduling strategy for the uplink TBFs which are active. This is also done by a round robin and priority scheduling strategy. After a block for each PDCH is scheduled the blocks are given to the PLL interface. Then the radio blocks which have been received in the block period are forwarded to the corresponding uplink or downlink FSMs. The task sequence for one block period is shown in Figure 7.14.

7.8.1.9 prBS_Clock

This process triggers the channel mapping process and the PLL interface each time a new time slot period begins.

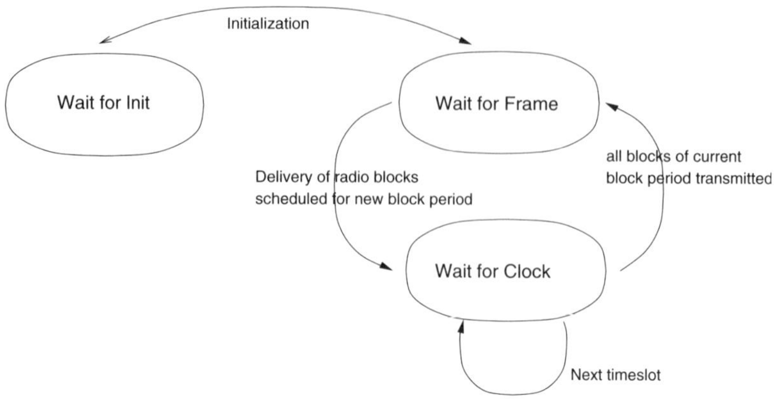

Figure 7.15: State machine of the process PLL interface

7.8.1.10 prBS_PLL_Interface

This process is responsible for the regular transmission of radio blocks over the radio channel. It is triggered by the clock process and transmits a new radio block if the related first burst period begins. When all blocks of the current block period are transmitted, the process waits for the next scheduled radio blocks which will be transmitted in the next block period. The state machine of the process is shown in Figure 7.15.

7.8.2 RLC/MAC in the Mobile Station

In Figure 7.16 the structure of the RLC/MAC layer in the base station is shown.

The processes `prMS_PLL_Interface` and `prMS_Clock` are identical. Since there is only one logical link, one uplink TBF and one downlink TBF associated with the mobile station, each process exists only once in the system initialized by the process `prMS_Config`. The LLC interface is not necessary since there is only one LLC entity and one RLC entity and thus no mapping function to be realized.

In the following the processes, which differ from the corresponding processes in the BS, are described.

prMS_RLC_Entity: This process is similar to the process RLC entity of the base station. It provides the same functions except the management of the ACK/NACK messages. This is only the task of the BS. A `PaDowACK/NACK` message has to be sent when the poll bit in the received data blocks is set. Uplink acknowledgement messages are sent from the BS automatically. As mentioned, the GPRSIM operates with open-ended TBFs. A countdown procedure for the uplink is not implemented. Only the last RLC data block in an uplink TBF is indicated by the *Countdown Value* (CV) zero.

prMS_DL_FSM: The downlink FSM manages the state sequence and behavior of the downlink TBF on the MS side. The state machine is shown in Figure 7.17.

After initialization the process waits for a `PaDowAss` message from the peer entity forwarded by the MS transceiver process. This corresponds to an idle state where no TBF exists. When this message arrives the TBF is established.

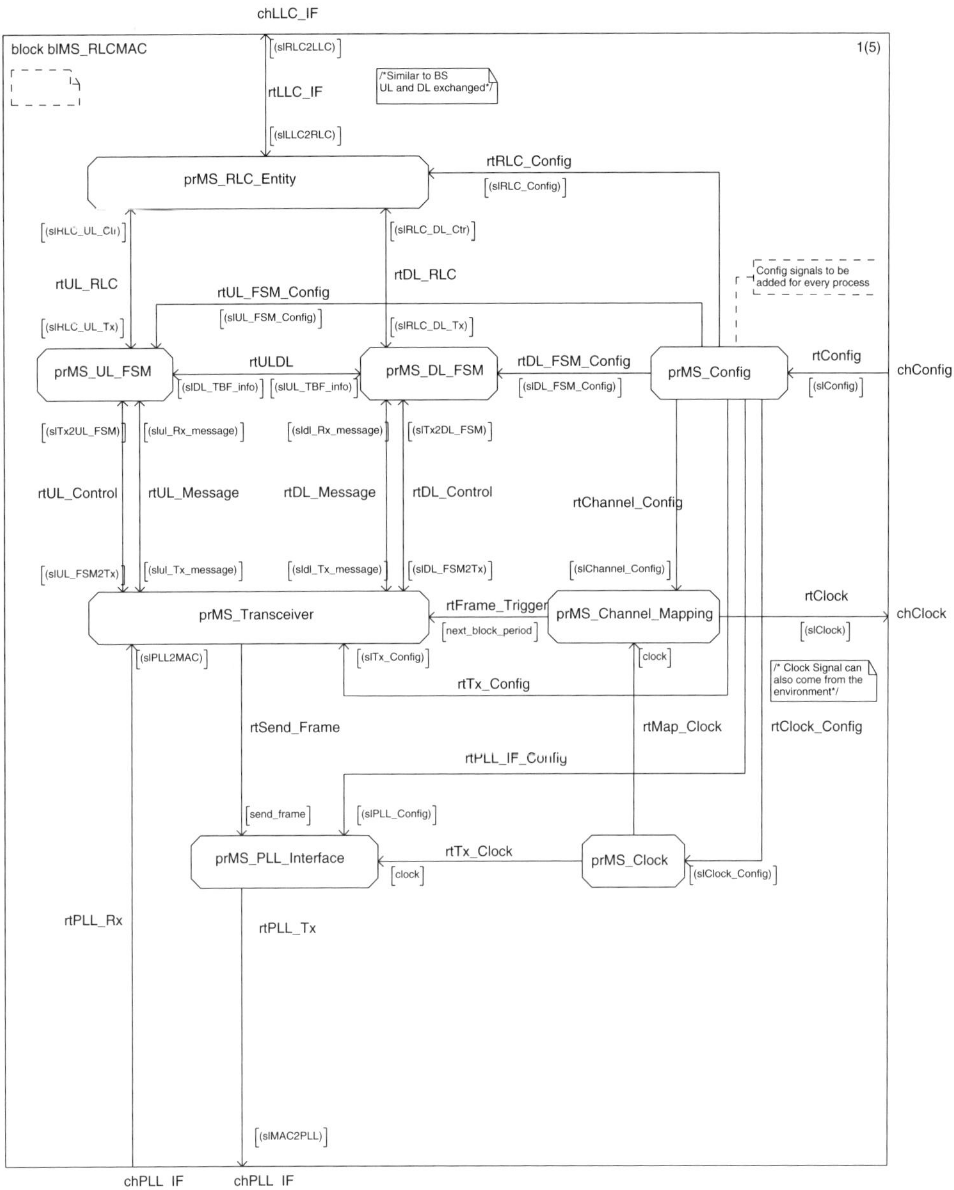

Figure 7.16: RLC/MAC mobile station architecture

In this state RLC/MAC blocks can be received and forwarded. If a Packet ACK/NACK message is polled by the peer entity, the RLC entity is asked for such a message and the process moves into the wait state. After the response from RLC it goes back to the former state. For every state it is ensured that a timer resets the process into the idle state if the process stays a certain time in a wait state without reception of a response signal. If a regular TBF end is indicated by the RLC entity, the TBF context is released after a delay in order

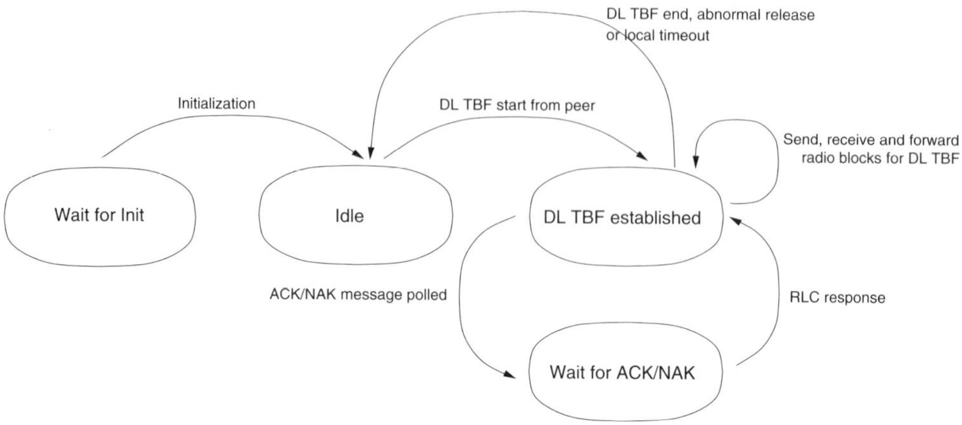

Figure 7.17: State machine of the process downlink FSM at the mobile station

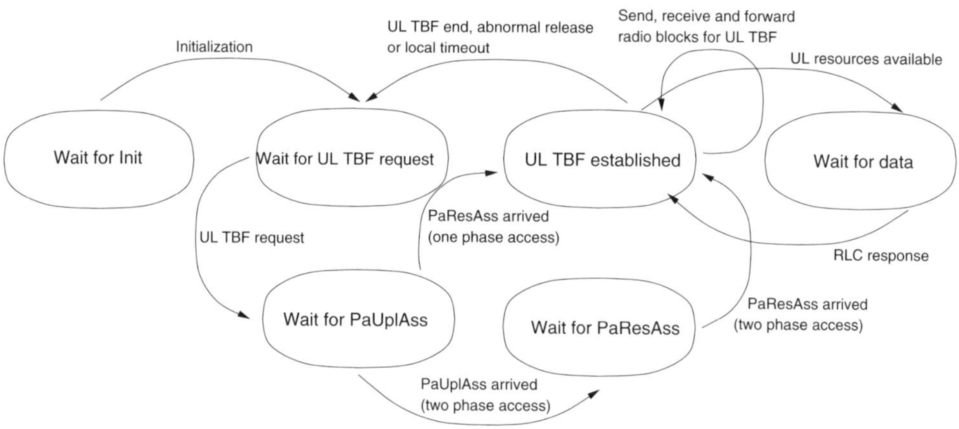

Figure 7.18: State machine of the process uplink FSM at the mobile station

to let the TBF open for new LLC data, which might be transmitted after a short inter-packet-train gap.

prMS_UL_FSM: As in the process downlink FSM the uplink FSM manages the state sequence and behavior of the uplink TBF at the mobile station. The state machine is shown in Figure 7.18.

After initialization the process waits for an uplink TBF initiation from the RLC entity. A `PaChaReq` message is coded and given to the transceiver process for transmission. The access mode, which can be either one-phase or two-phase access, is part of this block and is dependent on a configuration parameter. In one-phase access the process waits for a `PaUplAss` (`PaImmAss`) message and goes into the TBF Established state, when it arrives. In two-phase access the process waits for a `PaUplAss` message. When it arrives, a `PaResReq` message is sent to the transceiver and the process waits for a `PaUplAss` (`PaResAss`) message from the peer. After it is received the process goes into the state TBF Established and is ready for sending, receiving and forwarding of RLC/MAC blocks. If a

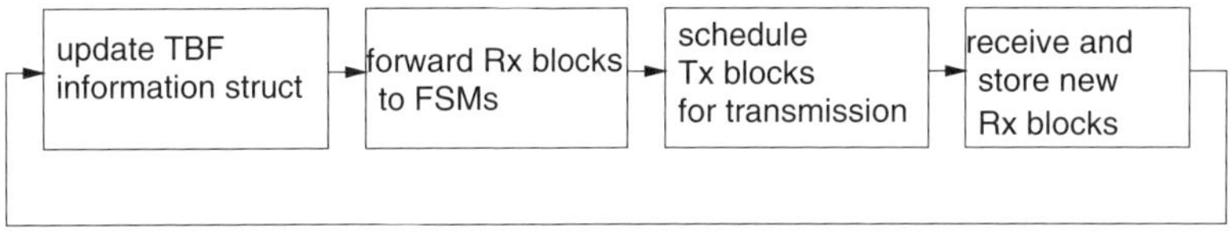

Figure 7.19: Task sequence in a block period at the mobile station

demand is made from the transceiver to transmit data, the process asks the RLC entity for a data block and the process waits for the RLC response. After this response a data block is forwarded to the transceiver or the transceiver is informed that there is no data to transmit. As in the downlink FSM, errors are resolved by resetting the process into the Idle state if a timeout occurs. If a regular TBF end is indicated by the RLC entity, the TBF context is released after a delay time in order to let the TBF open for new LLC data which has to be transmitted after a short inter-train gap.

prMS_Channel_Mapping: This process models the multiframe structure of (E)GPRS for the mobile station. Since logical channels are indicated by the USF, this process has the sole task of triggering the transceiver each new block period and of indicating the multiframe structure to the environment.

prMS_Transceiver: The principal data structures in the transceiver are

- A TBF information table, which stores the information of this logical link.
- An USF array, which stores the USF indicated by the peer entity in the last block period.
- Buffers for received blocks, which have to be processed and forwarded.

The receiving procedure is similar to that for the transceiver at the base station. Radio blocks are received and the header information of these blocks is examined. It extracts the TBF that is addressed. If the station is addressed by this TBF identifier, the process updates the data structures depending on the message type and its parameters and stores the blocks in the receive buffer. Additionally the USF for the related PDCH is stored in the USF array. Stored blocks are forwarded later to the uplink or downlink FSM. If a resource assignment message arrives, resource allocation information has to be stored in the TBF information structure. If the station is polled for a `PaResReq` or a `PaACK/NACK` message the blocks are requested from the higher processes and directly scheduled for transmission at the correct time slot. The transmission of RLC/MAC blocks works as follows. When a new block period is indicated by the channel mapping process, the PDCHs that are allocated for transmission are examined. If a poll command was indicated by the peer, a `PaResAss` or a `PaACK/NACK` message is demanded from the higher layers. If the poll bit was not set and the USF is set on the value `FREE` then a `PaChaReq` message may be sent. If the indicated USF has another value, its own USF entry for the regarded PDCH is compared with this value. If it is equal, the higher processes are requested for data to transmit. At the end of the block period the RLC/-MAC blocks that it was decided to transmit in the related slots are given to the PLL interface. The task sequence for one block period is shown in Figure 7.19.

7.9 Transceiver

The transceiver modules have the task of forwarding the radio blocks which are given from the RLC/MAC layer to the peer entity or of forwarding the radio blocks which are received from the peer entity to the RLC/MAC layer. An error flag which is part of the radio block data structure is set if a collision of packet channel request blocks of several MSs is detected or if the channel model generates a block error. Additionally the delay which corresponds to the duration of the block transmission on the channel interface is realized here.

7.9.1 Channel Models

The channel models define the state of the radio channel and the resulting block errors for uplink and downlink radio blocks. The GPRSIM provides three channel models.

7.9.1.1 Channel Model 0 (error-free channel)

This channel model assumes an error-free channel. It can be used for GPRS and EGPRS simulations. Only when a collision of packet channel request blocks of several MSs is detected are the corresponding blocks rejected by setting the error flag.

7.9.1.2 Channel Model 1 (GPRS Channel Model)

Channel model 1 is a model for pure GPRS simulations. Mobile stations not supporting EGPRS may be active. Input of the error model is a fixed C/I ratio and a fixed *Coding Scheme* (CS) which can be selected before simulation starts. Control block errors are not regarded. The look-up function giving a *Block Error Probability* (BLEP) for every C/I ratio is taken from link level simulations [59, 114, 115]. A block error occurs if one or more bits in a coding block (1 block = 4 bursts) is erroneous after decoding. Perfect error detection is assumed. The TU3 (Typical Urban) channel model in GSM Rec. 05.05 [20] is used in all link level simulations reported later. Frequency hopping is not used. The channel fades according to 3 km/h on a 900 MHz channel during the burst. The results are obtained without antenna diversity.

In Figure 7.20 the BLEP versus C/I that formed the basis for the look-up function used in channel model 1 is shown. When radio blocks are exchanged between the transceiver modules, the channel model determines whether the arrived block is correctly received at the receiver or not.

7.9.1.3 Channel Model 2 (EGPRS Channel Model)

This channel model is usable for both EGPRS and GPRS. Figure 7.21 shows a block diagram of the model.

To generate block errors a set of mapping curves is used. These curves are gained from link level simulations and allow the mapping of C/I values to the corresponding block error probabilities for every radio block as described in Section 7.9.1.2.

Beside the BLEP value channel model 2 determines the optimal MCS for the actual channel quality (see Figure 7.21).

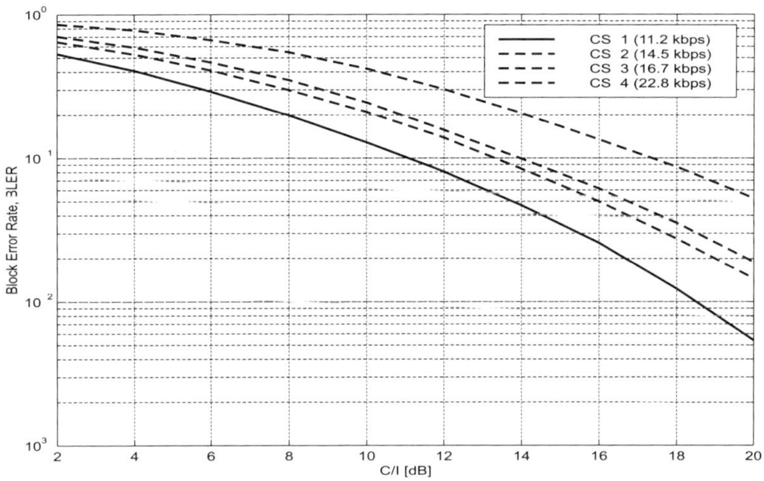

Figure 7.20: BLEP over C/I reference function used for channel model 1 (GPRS)

Channel Model 2 for EGPRS The value of the C/I ratio on burst basis is mapped onto a generic burst quality measure Q, which is directly related to the BLEP. The mapping from the C/I ratio to Q is done according to the function given in Figure 7.22.

In a second step the values Q_{mean} and Q_{std} are calculated for every data block. The number of bursts per data block depend on the selected MCS. If the MS operates with MCS-1–MCS-6 every data block is interleaved over four bursts. For MCS-7, MCS-8, and MCS-9 one data block is interleaved over two bursts. If IR is used, the Q values of the previous transmissions are considered. The Q_{mean} and Q_{std} values are then mapped to the BLEP using appropriate mapping curves, which are obtained from link level simulations [19]. For each MCS one mapping table is implemented in the simulator. As an example, two of these curves are shown in Figure 7.23.

Channel Model 2 for GPRS For GPRS transmissions the values of the C/I ratio on burst basis are used to calculate an average C/I. Since in GPRS no IR is possible, no standard deviation has to be taken into account. The average C/I is mapped directly onto a BLEP. The look-up functions of the BLEP and probability are derived from effects on the physical layer, such as fast fading, frequency hopping, channel coding and equalization.

As in Section 7.9.1.2 the reference function of BLEP versus C/I, which is taken from link level simulations, is used for GPRS in channel model 2. The parameters for the (E)GPRS channel model 2 are listed in Table 7.1.

7.9.1.4 Channel Model 3 (Gilbert-Elliot Model)

Channel model 3 is an enhanced channel model 1 for pure GPRS simulations. This channel model is based on the *Gilbert model* [61], which is in turn based on a two-state *Markov chain*. This model offers the possibility of simulating burst errors. Figure 7.24 illustrates the model and the state transitions.

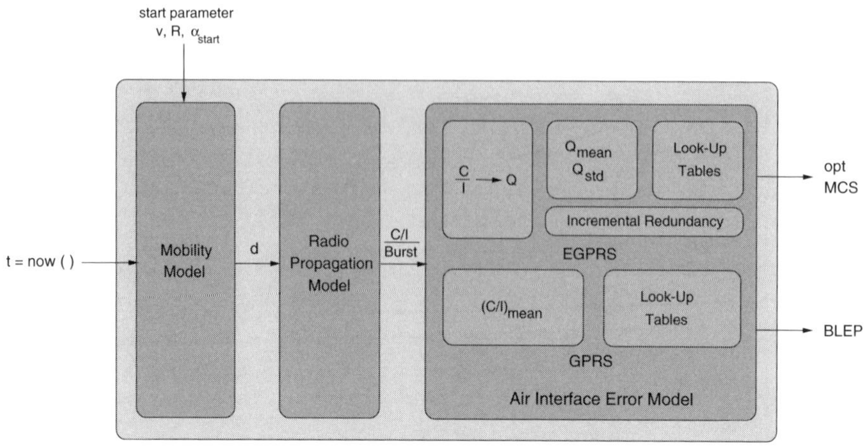

Figure 7.21: Block diagram of channel model 2

Table 7.1: EGPRS channel model parameters

Parameters	Default values
Velocity of the MS	100 km/h, 6 km/h
Cell radius	3 km, 300 m
Cluster size	3, 7, 9
C/I variance (shadowing)	7
Path loss parameter γ	3.7

To simulate burst errors, the states B and G must tend to persist, which means that the transition probabilities p_{GB}, p_{BG} are small. The Gilbert model was extended by *Elliott* [51], in so far as the probability for errors in the state *GOOD* is no longer zero. In this way there is the opportunity of simulating slight background noise in the good state.

7.9.2 Mobility Model

The mobility model determines the actual distance between MS and BS within every time step of the simulation. The distance is used by the radio propagation model, together with statistical behavior assumptions of the channel and the transmission activities of MS and BS, to calculate the signal-to-interference ratio on the receiving side. Within the *air interface error model* the BLEP is derived and the radio block is set to either error-free or defective.

Two different mobility models are implemented.

7.9.2.1 Secant Mobility Model

This mobility model determines the distance between BTS and MS in every time step (a radio block period) of the simulation. During start-up time a randomly located start point on the cell border (determined by a start angle α_{start}) and a given velocity of movement are assigned to every MS.

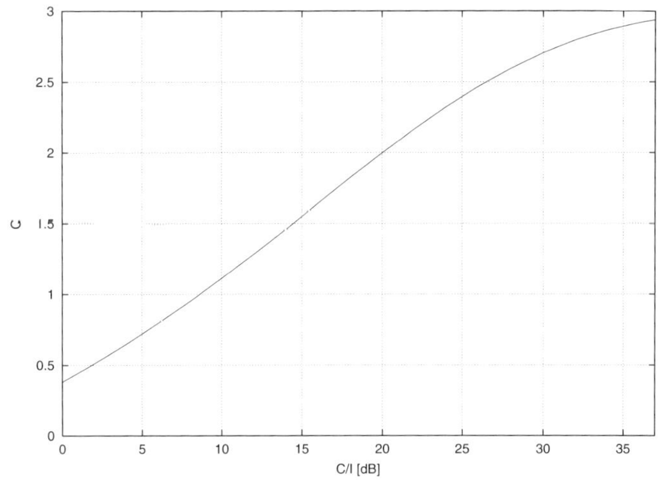

Figure 7.22: C/I to Q mapping used for channel model 2 (EGPRS)

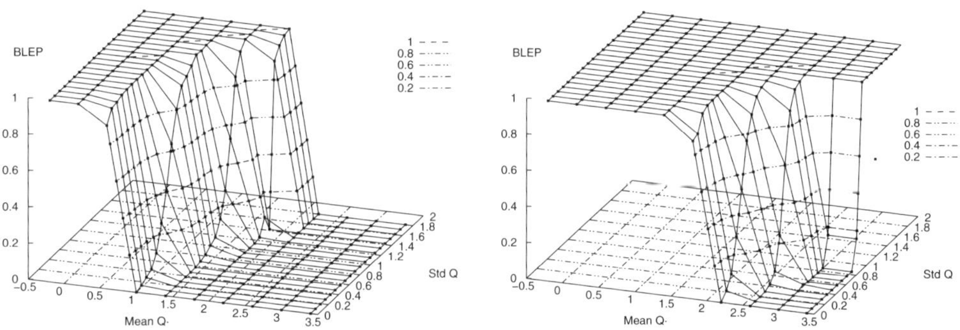

Figure 7.23: Mapping of Q_{mean} and Q_{std} to BLEP

When the simulation starts, the MS starts moving with the assigned velocity on a horizontal line through the circular cell with radius R. If a mobile station reaches the cell boundary, it starts on the cell boundary and moves again on a straight line through the cell.

Figure 7.25 illustrates how a single MS moves through the cell, reaches the cell boundary and starts again moving from a random location on the cell boundary. This model is a rather simple model and does not take into account the fact that mobile stations could change their direction or their velocity. But for the basic effects of LQC, examined within the example studies presented later, it is sufficient. In Figure 7.26 the actual distance between one MS and the BTS is shown over the simulation time where the MS has a velocity of 100 km/h.

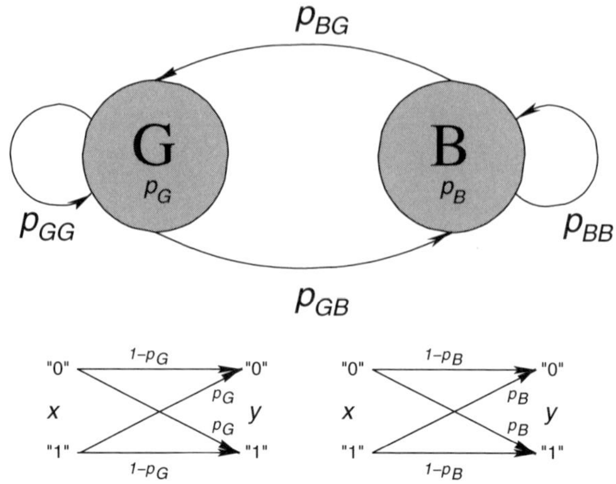

Figure 7.24: The Gilbert-Elliott channel model

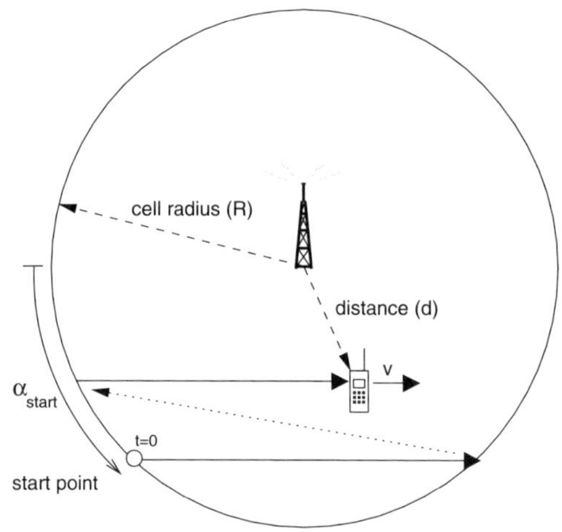

Figure 7.25: Mobility model

7.9.2.2 Brownian Mobility Model

This is an enhanced model to simulate movement of mobile stations more realistically. It is based on the *Brownian motion* of molecules [81], whose speed and direction depend on the collisions among each other. This mobility model determines the distance between BTS and MS in every time step (a radio block period) of the simulation, like the standard model. During start-up time a randomly located start point in the cell and a given velocity of movement are assigned to every MS. When the simulation starts, the MS starts moving in the randomly defined direction. In every time step, there exists a chance of changing the direction for every MS. If a mobile station reaches the cell boundary, it restarts at a randomly defined position in the cell. Figure 7.28 illustrates the movement of a single MS in the cell.

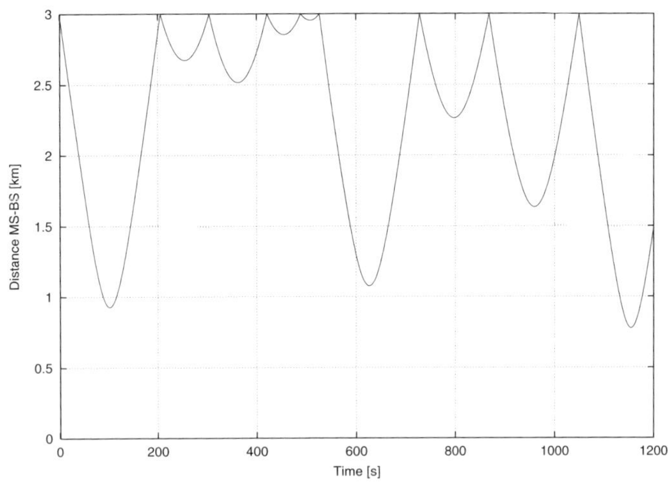

Figure 7.26: Distance between MS and BS using the Secant Mobility Model

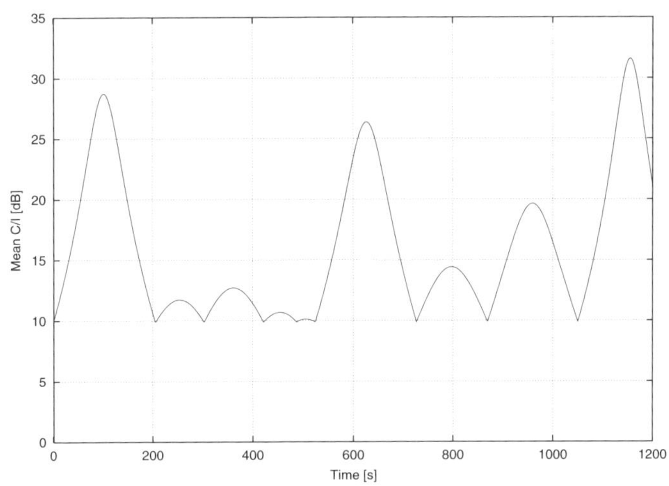

Figure 7.27: Mean C/I ratio over time, using the Secant Mobility Model

7.9.3 Radio Propagation Model

The influence of radio propagation is estimated in two steps. First, the C/I ratio is calculated using:

$$\frac{C}{I}[\text{dB}] = 10 \cdot log\left(\frac{d^{-\gamma}}{6 \cdot D^{-\gamma}}\right) \text{ with } D = R \cdot \sqrt{3 \cdot C} \tag{7.1}$$

where d is the distance between MS and BS, R the cell radius, C the cluster size, D the co-channel distance and γ the path-loss factor.

Since the long-term fading mean-value changes relatively slowly with the distance between MS and BTS, the distance-dependent calculation of the pathloss-related C/I

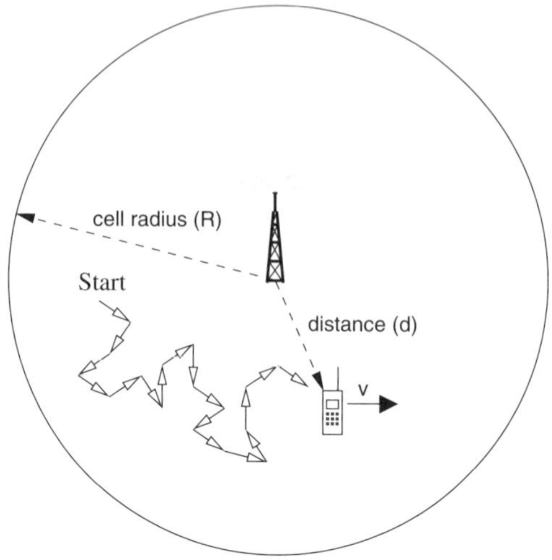

Figure 7.28: Brownian motion

value is carried out only once per radio block. The actual distance between MS and BTS required for this calculation is gained from the mobility model.

In a second step, long-term fading through signal shadowing is modeled by adding a normally-distributed random variable with a mean of zero and a given variance to the mean pathloss value. The calculation of the impact of shadow fading on C/I is performed on a burst basis (four times per radio block). Figure 7.27 illustrates how the mean C/I ratio alters over time if the distance between MS and BTS changes as shown in Figure 7.26.

Short-term fading is already included in the link level simulations and needs therefore no further consideration at the system level.

7.10 Input-Output Behavior

7.10.1 Initialization

The initialization is performed comprising the following steps:

- Read the simulation parameters from the file *.sim_defaults*.
- Read the Look-Up Tables for the channel model.
- Create the several class instances.
- Send the *Start Events* to the generator.
- Create and initialize the SDL systems.
- Start the scheduler.

The parameter file `.sim_defaults` is read at the start of the simulation, so simulation parameters can be changed without recompiling the program.

7.10.2 Output/Statistics

After a simulation run, statistics are produced in a results file. *Inter alia* the following statistics (mean, variance, quantiles and distribution functions) in uplink and downlink and for all different application and subscriber classes are available:

- IP user throughput
- IP datagram delay
- Application response time
- Session abortion rate (session blocking)
- CS call blocking rate
- IP cell throughput
- PDCH utilization
- Assigned PDCHs

7.10.3 Graphical Presentation

To make the protocols' functionality transparent the simulation flow can be presented with a graphical user interface. The tool *Graphical Interactive Simulation result Tool* (GIST) [31] is utilized for this. GIST is an independent process called by the main program. It offers the simulation an interface to exchange data with this tool. Navigation over the simulation time can be done by using the buttons used in a video recorder. They provide the control of simulation sequences with a configurable speed.

The GIST graphic (see Figure 7.29) shows a 52-multiframe. For every time slot the bursts on uplink and downlink and the USF are pictured. Also the LLC protocol sequence is represented by LLC message sequence charts.

7.11 Web Interface

The web interface was created to simplify the simulator handling and to enable start and evaluation of simulation runs over the Internet.

After authentication, the user arrives at the start page of the simulator, where he can decide between the menu points simulation parameters and simulation progress. At the menu simulation parameters the user can change a certain number of parameters and start a new simulation after confirmation (see Figure 7.30). It is also possible to use a help function that explains the existing parameters. To avoid conflicts between several users and to guarantee discretion between users from different companies, for each user a new domain is created. At the menu simulation progress the user can supervise the progress of a current simulation run and read the temporary evaluation files that are created at specified intervals and at the end of the simulation.

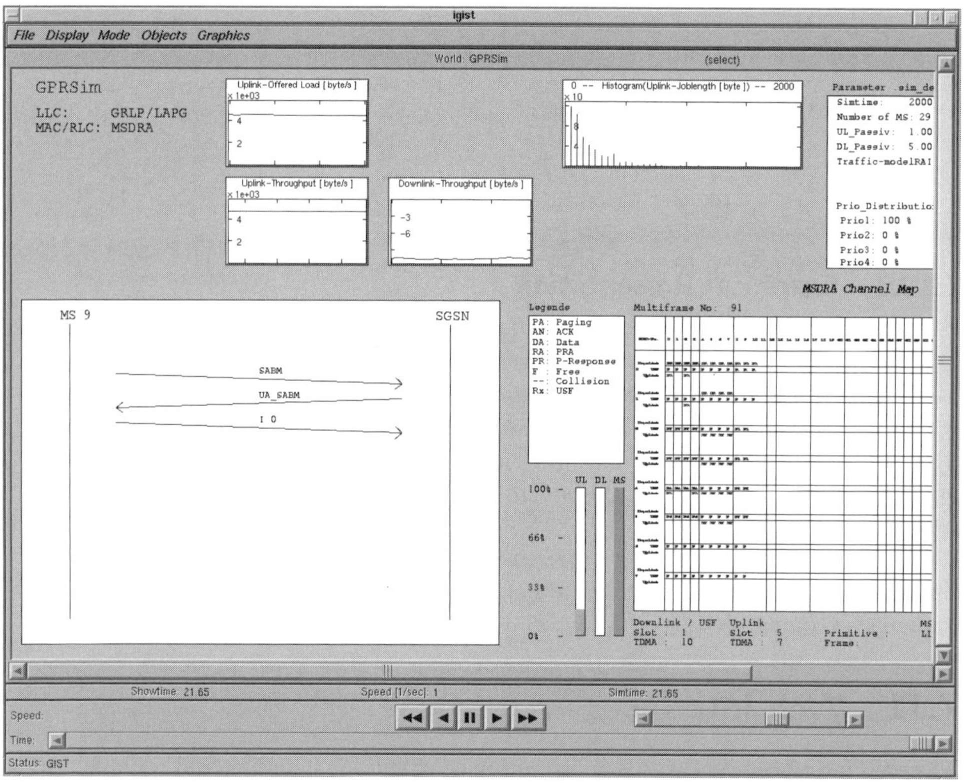

Figure 7.29: Graphical GIST presentation of the MAC and LLC transfer

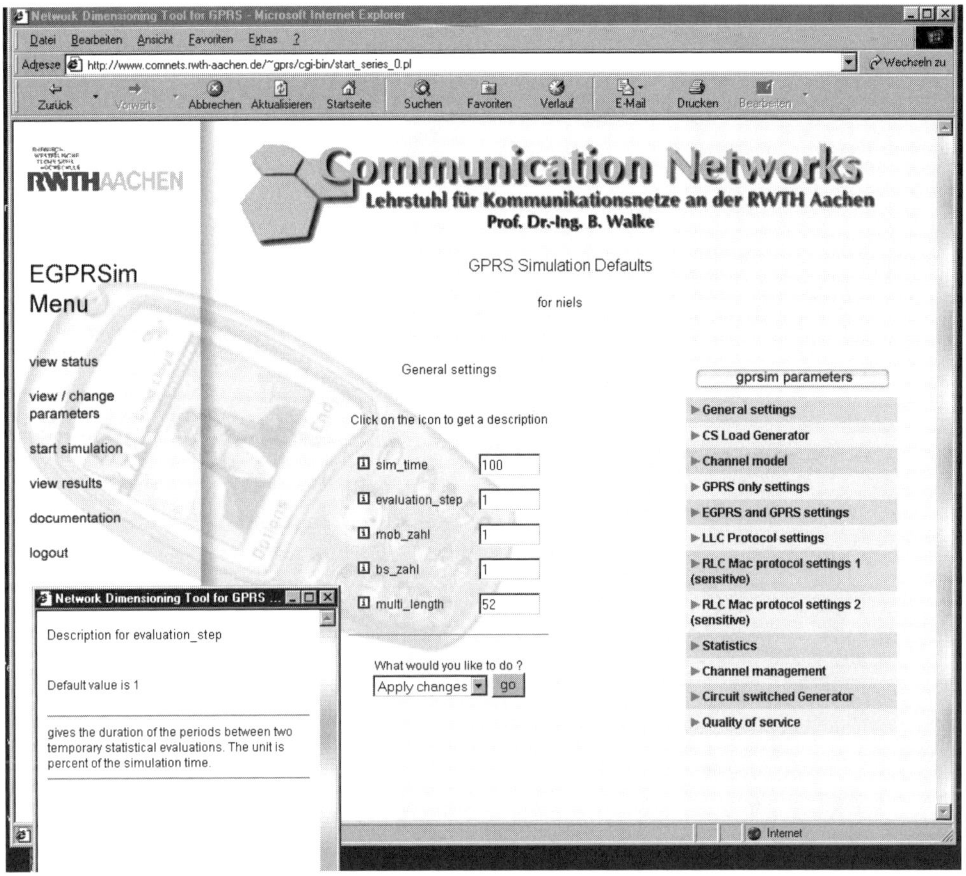

Figure 7.30: The GPRSim web interface

8
Traffic Performance of GPRS and EGPRS

This chapter presents end-to-end traffic performance results based on the GPRSim (see Chapter 7) for different traffic mixes of typical applications in GPRS and EGPRS networks.

First the traffic performance of Internet applications over GPRS is presented. Both fixed and on-demand PDCHs are regarded. To compare the packet data services GPRS and EGPRS the effect of the new modulation and coding schemes and the LQC functions LA and IR are outlined.

Next, different applications and traffic mixes typical for the introduction and the evolution of GPRS and EGPRS are examined. After regarding WWW, e-mail, WAP and Streaming applications separately, traffic mixes are examined. One traffic mix is representative for a GPRS evolution scenario and one for an EDGE introduction scenario. The first one is composed of WAP, WWW and e-mail; the second is defined by Streaming, WWW and e-mail.

Finally the performance gain through QoS management differentiating both subscribers and applications in GPRS networks is presented. Both the effect of *Connection Admission Control* (CAC) and scheduling in the GPRS radio access network and of service differentiation in the GPRS core network are examined.

8.1 General Simulation Parameter Settings

The cell configuration is given by the number of *Packet Data Channels* (PDCHs) permanently available or available on demand for GPRS or EGPRS.

An RLC/MAC block error rate of 13.5% has been assumed throughout the GPRS simulations corresponding to a C/I of 12 dB. This parameter value corresponds to a typical radio coverage planning goal widely used by operators for GSM networks. CS-2 is used as the coding scheme for user data, which is a relatively robust coding scheme. Therefore, capacity limitations caused by co-channel interference in high load conditions can be neglected. We have found from sample simulations of multi-cell scenarios where co-channel cells were included in the model that the results presented from single cell simulations in fact can be trusted as being representative [57].

For EGPRS the channel conditions are determined by the cell and cluster size that are the basis for the C/I calculation as described in Section 7.9.3. Cluster size 3, 7 and 9, a cell size with a radius of 300 and 3000 meters and a velocity of 100 and 6 km/h have been regarded. The secant mobility model is used in the EGPRS simulations (see Section 7.9.2.1).

Mobile stations with the multislot capability of 1 and 4 in uplink and 2 and 4 in downlink are discussed. These are typical values for a first phase of GPRS introduction. LLC and RLC/MAC are operating in acknowledged mode for WWW, e-mail,

WAP and Streaming. The MAC protocol instances in the simulations are operating with three random access subchannels per 52-multiframe. In the simulations without QoS management, all conventional MAC requests have the radio priority level 1 and are scheduled with a Round Robin strategy with 10 radio blocks per TBF per Round Robin cycle. LLC has a window size of 16 frames. TCP/IP header compression in SNDCP is performed. The maximum IP datagram size is set to 1500 byte for WAP and 552 byte for the TCP-based applications. In the Internet stack for WWW and e-mail TCP is operating with a maximum congestion window size of 8 Kbyte. The transmission delay in the core network and external networks, i.e., the public Internet, is neglected (except in Section 8.6.3.2), since it is assumed that the servers are located in the operator's domain and the core network is well dimensioned. Since the high round-trip time in GPRS networks is mainly caused by *Temporary Block Flow* (TBF) establishment procedures at the air interface, the delay in well-dimensioned IP subnetworks does not have a great effect on the end-to-end performance.

In order to perform network capacity planning the system behavior during the *busy hour* has to be considered. The traffic load generated is characterized by the number of Internet sessions running or by the offered traffic measured in kbit/s per radio cell. The traffic characteristics are related to the traffic models introduced in Chapter 6. While for WWW sessions the adapted Mosaic model is used (see Section 6.1.1), e-mail, WAP and Streaming sessions are characterized by the models in Section 6.2.2, Section 6.4.4 and Section 6.5.1.3, respectively. The inactive period between two sessions is negative-exponentially distributed for all kinds of (E)GPRS sessions with the mean value of 60 seconds. In the following, mobile stations running (E)GPRS sessions with an average 60 seconds inter-arrival times of sessions are called 'active users' or 'active stations'.

Parameters that differ between the following scenarios will be explicitly defined at the beginning of the related sections.

8.2 Performance and Capacity Measures

To characterize the traffic performance of GPRS several performance and system measures are defined in the following. For the different applications different critical performance measures will be regarded, since different performance characteristics are required for download-oriented, transaction-oriented and real-time applications, respectively. While for the download-oriented and transaction-oriented applications mean values represent the user-perceived performance [80], variance and distribution functions should be regarded for real-time applications.

Mean IP throughput per user is the downlink IP throughput measured during transmission periods, e.g., the download period of a single object of a web page. This is an important QoS parameter from a user's point of view. This measure is evaluated statistically by counting the amount of IP bytes transmitted in each TDMA frame period for each user, if a packet train is running. Thus, the throughput is not averaged over inactive periods. The number of IP bytes transmitted divided by the TDMA frame duration represents a simulation sample value in the evaluation sequence. At the end of the simulation the mean throughput is calculated from this evaluation sequence.

Mean application response time is the difference between the time when a user is requesting a web page, a WAP deck or an e-mail and the time when it is completely received.

Mean IP datagram delay is the end-to-end delay of IP datagrams evaluated by means of time stamps given to the datagrams, when the IP layer performs an SNDCP data request for transmission. When the datagram arrives at the receiver, the difference between the actual time and the time stamp value is calculated as a sample of the respective evaluation sequence.

95-percentile of the mean IP datagram delay is the 95-percentile of the previous measure.

Abortion rate is the number of aborted sessions due to bad performance normalized to the total number of sessions.

Mean throughput per cell is also called system throughput and is calculated from the total IP data transmitted on all PDCHs of the regarded radio cell and for all users during the whole simulation duration, divided by the simulation duration. Since a loss of IP datagrams over fixed subnetworks is not modeled, this parameter equals the offered IP traffic in the radio cell.

PDCH utilization is the number of MAC blocks utilized for MAC data and control blocks normalized to the sum of data, control and idle blocks. Thus existing capacity reserves in the scenario under consideration can be seen from this measure.

8.3 Simulation Results for GPRS

In GPRS networks a radio cell may allocate resources by means of one or several physical radio channels in order to support the GPRS traffic. The channels shared by the GPRS mobile stations for packet transmission are taken from the common pool of GSM physical channels available for speech and packet data communication in the radio cell. The allocation of physical channels to circuit-switched services and to the GPRS packet-switched service is done dynamically according to the "capacity on demand" principle [17], whereby the operator can decide to permanently or temporarily dedicate physical channels for the GPRS. In this context GSM physical channels that are allocated permanently for GPRS are called *fixed Packet Data Channels (PDCHs)*, while channels that are allocated temporarily for GPRS are called *on-demand PDCHs* (see Section 2.2.5.1).

In the following subsections simulation results for scenarios with fixed and on-demand PDCHs regarding a traffic mix of WWW/e-mail (see Section 6.1.1, Section 6.2.2 and Section 7.2) are discussed. The probabilities for a WWW and an e-mail session are 30% and 70%, respectively.

8.3.1 Fixed PDCH Scenarios

To give a first overview on the traffic performance of Internet applications over GPRS, simulation results for scenarios with a fixed number of PDCHs are presented in this section.

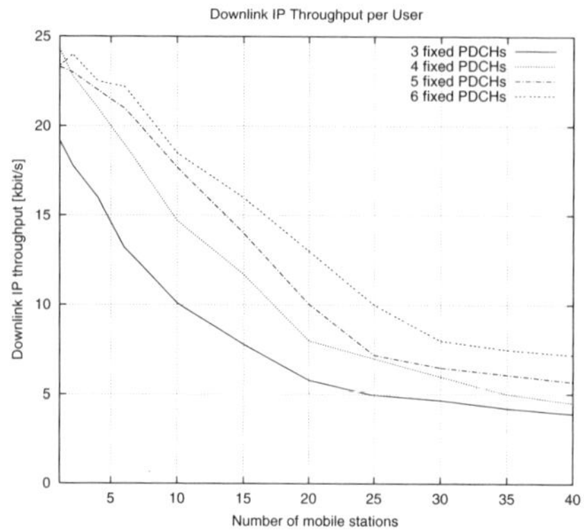

(a) Downlink IP throughput per user

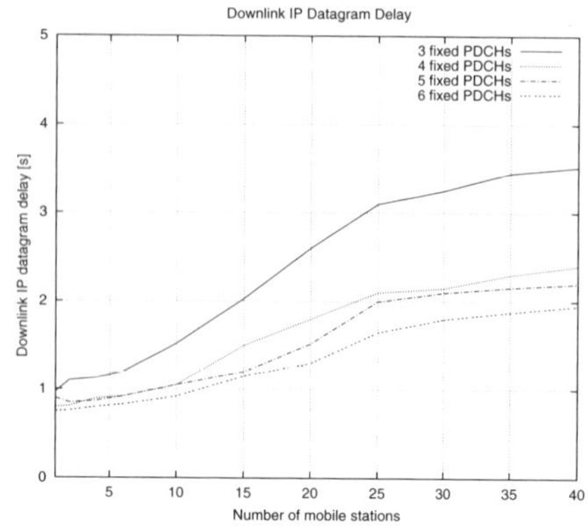

(b) Downlink IP datagram delay

Figure 8.1: Performance measures for different numbers of fixed PDCHs

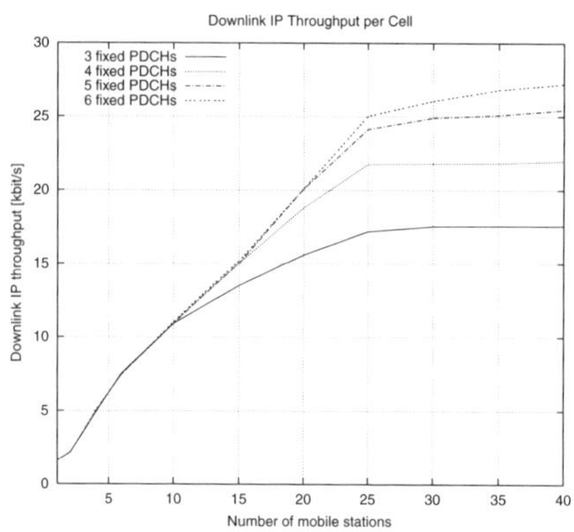

(a) Downlink IP system throughput

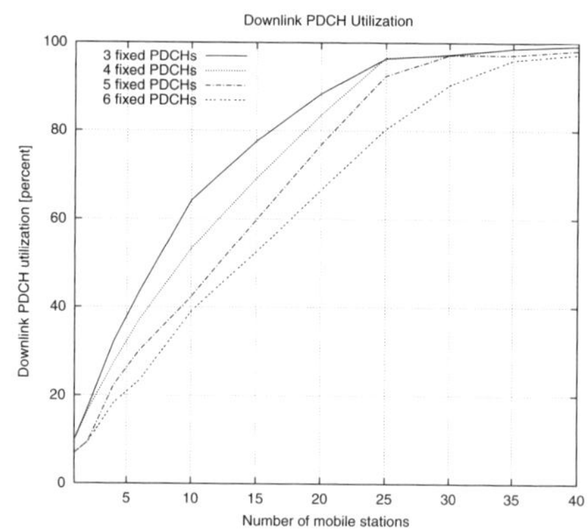

(b) Downlink PDCH utilization

Figure 8.2: System measures for different numbers of fixed PDCHs

Figure 8.1 and Figure 8.2 show the performance and system measures over the number of mobile stations with the multislot capability of four uplink and downlink slots in the cell for a different number of PDCHs available for GPRS. The maximum downlink IP throughput per user during a transmission is about 24 kbit/s and decreases with the number of mobile stations. The delay increases from about 1 second to about 3 seconds per IP packet, when 25 stations are active. The delay increases from about 1 second with a small number of active stations to about 3 seconds per IP datagram, when 25 stations are active. With 20 to 25 active mobile stations the system reaches saturation, which can also be seen from the results presented for the PDCH utilization and the total throughput per cell (see Figure 8.2).

Figure 8.3(a), Figure 8.3(b) and Figure 8.4(a) are based on the same simulations, but show performance parameters plotted over the offered IP traffic.

The downlink IP throughput per user during transmission decreases nearly linearly with an increasing offered IP traffic. For three fixed PDCHs the IP datagram delay reaches a pole with an offered IP traffic of around 20 kbit/s. The poles for config-

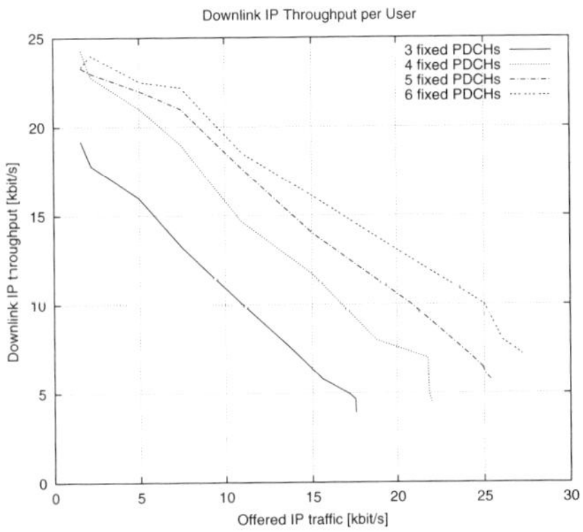

(a) Downlink IP throughput per user

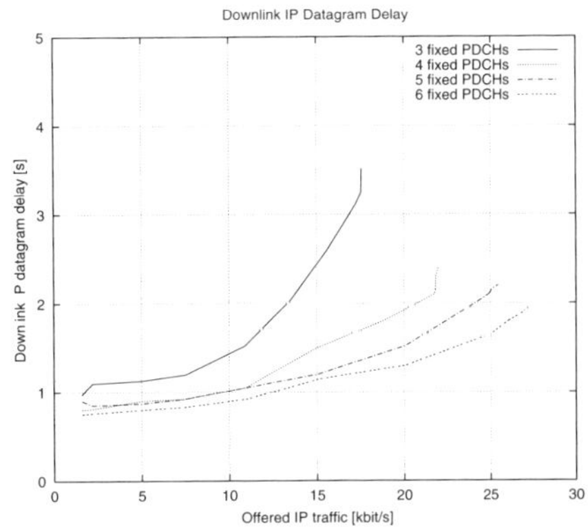

(b) Downlink IP datagram delay

Figure 8.3: Performance measures over the offered traffic

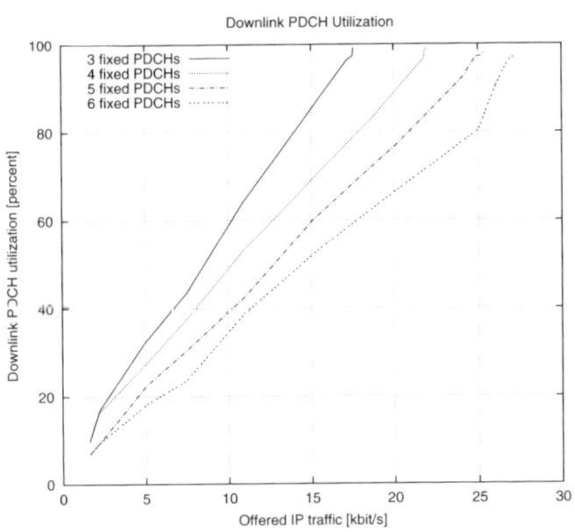

(a) Downlink PDCH utilization

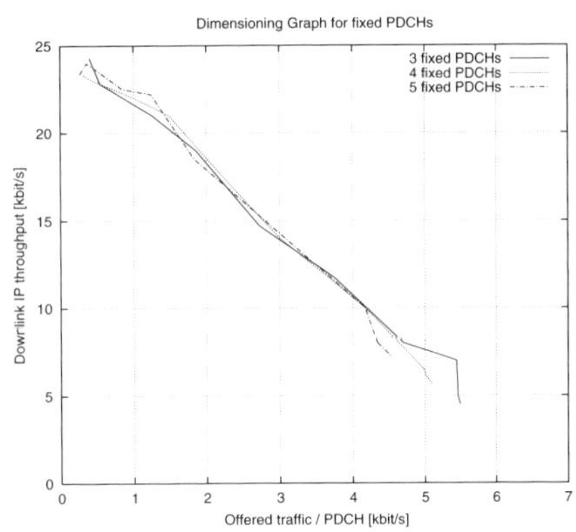

(b) Dimensioning graph for fixed PDCHs

Figure 8.4: PDCH utilization and dimensioning graph

urations with more than three PDCHs are not reached in the simulations, since the capacity limit is not reached for up to 25 stations studied.

The downlink IP throughput per user during transmission decreases almost linearly with increasing offered IP traffic. For three fixed PDCHs the IP packet delay reaches a pole with an offered IP traffic of around 20 kbit/s. The poles for configurations with more than three PDCHs are not reached in the simulations, since the capacity limit is not reached for up to 25 stations studied.

We base our traffic engineering rules on a realistic model with realistic parameter settings, while the rules themselves should be applicable only based on the user number and traffic volume during the busy hour. In planning scenarios, typically hundreds of potential users per cell are regarded that can be active or inactive offering traffic volumes in the order of several kbyte/h. Thus, we derive our traffic engineering rules from the simulation results by mapping a large number of potential active or inactive users generating a small amount of traffic on our simulation scenarios, where only the active users are regarded offering the same aggregate offered traffic as all potential users per cell in the planning scenario.

What is needed for engineering the number of fixed PDCHs per cell under a given traffic load is a reliable relation between the mean downlink IP throughput requested per user and the offered traffic per PDCH. This relation can be gained by normalizing the offered traffic of Figure 8.1(a) to the number of fixed PDCHs used as shown in Figure 8.4(b). Thereby, the curves for more than three PDCHs are transformed into only one curve, the dimensioning graph that we propose for traffic engineering of GPRS cells with a fixed number of PDCHs, by means of the following five steps:

Step 1: define the desired QoS

Step 2: estimate the number of users per cell

Step 3: define the offered IP traffic per user and calculate the offered traffic per cell

Step 4: determine the acceptable traffic per PDCH with the desired QoS from the dimensioning graph in Figure 8.4(b)

Step 5: calculate the number of PDCHs needed from:

$$\text{PDCH} = \frac{\text{estimated traffic}}{\text{acceptable traffic per PDCH}} \tag{8.1}$$

To visualize this procedure, the following example is given. With a desired downlink IP throughput of 12.5 kbit/s, a given offered traffic per user of 54 kbyte/h, and 100 users per radio cell the needed number of PDCHs can be estimated by the following four steps:

Step 1: The mean downlink IP throughput of 12.5 kbit/s is the QoS requirement.

Step 2: Mean expected number of users = 100

Step 3: Offered traffic/user = 54 kbyte/h = 0.12 kbit/s

total offered traffic = 100 · 0.12 kbit/s = 12 kbit/s

Step 4: From Figure 8.4(b) with the QoS requirement of 12.5 kbit/s

acceptable traffic/PDCH = 3.5 kbit/s/PDCH

needed PDCHs = 3.4

Since an integer value of PDCHs has to be aimed at, in this example four fixed PDCHs would be needed.

8.3.2 Effect of the Multislot Capability

The simulation results shown in Figure 8.5 and Figure 8.6 visualize the difference between the multislot capabilities of two and four in the uplink and downlink directions. The scenario is modeled according to the previous section with four fixed PDCHs.

The maximum downlink IP throughput in the simulation with multislot capability 2 is about 60% of the value reached in the simulation with multislot capability 4. The same performance is reached for the different capabilities in situations with higher traffic load, since the PDCH utilization allows no higher throughput.

The results in Figure 8.7 are based on the same simulations but show the performance parameters over the offered IP traffic.

As depicted in the previous graphs, the downlink IP throughput is equal for the different multislot capabilities at 15 kbit/s, for the scenarios considered. The IP delay has a pole for an offered IP traffic of 25 kbit/s not depending on the selected multislot class.

8.3.3 On-demand PDCH Scenarios

Since GSM networks are dimensioned for a low blocking probability P_b for speech calls, the probability that a number of channels of the trunk of speech channels are unused is high. On the other hand in GPRS introduction scenarios the offered data traffic will be relatively low. As a result the use of on-demand PDCHs that are completely shared between voice and data services leads to an especially efficient radio resource utilization [106].

An example scenario with three *Transceivers* (TRXs) (21 TCHs) is discussed. For example, an offered voice traffic of 11.85 Erlang leads to $P_b = 0.5\%$ after the Erlang-B formula. The number of unused GSM TCHs that are available on average for GPRS is approximately eight [106].

In Figure 8.8 and Figure 8.9 the performance and system measures for different coexisting CS speech traffic values, corresponding to P_b values of 0.5, 2 and 10%, are shown.

While for a P_b of 0.5% the mean throughput performance (see Figure 8.8(a)) is similar to the performance with eight fixed PDCHs (see Figure 8.1(a)), the throughput decreases much faster regarding coexisting CS traffic corresponding to a P_b of 10%. Here the throughput passes below 10 kbit/s already with six GPRS stations offering Internet traffic. The mean throughput performance for a P_b of 2% decreases only by a maximum of 19% compared to coexisting CS traffic corresponding to $P_b = 0.5\%$.

In Figure 8.10 the performance measures in on-demand scenarios plotted over the offered GPRS traffic are shown. For Internet applications traffic engineering for on-demand configurations can be done by taking a dimensioning graph based on GPRS performance results for an existing cell scenario with on-demand PDCHs as the basis and by defining the acceptable coexisting CS traffic (corresponding to a given P_b value) so that the GPRS performance for a given offered IP traffic can be guaranteed. As an example, based on the three TRX scenario (see Figure 8.10(a)), this can be done according to the following five steps with the same example values as in Section 8.3.1:

Step 1: Estimate P_b for the related TRX scenario for the CS traffic, which is expected for the scenario considered. Here $P_b = 1.5\%$ is assumed.

Step 2: Mean expected number of users = 100

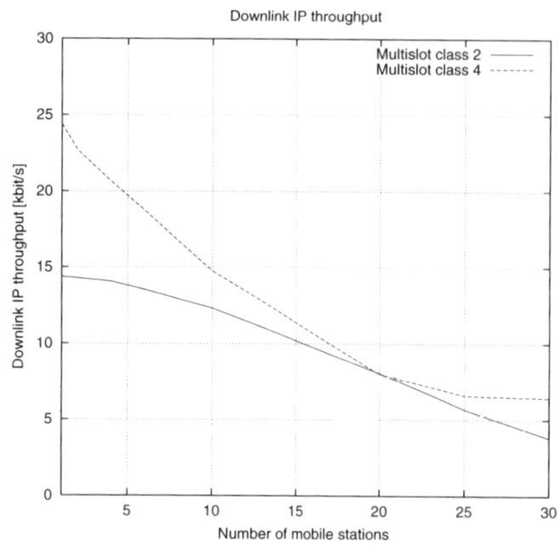

(a) Downlink IP throughput per user

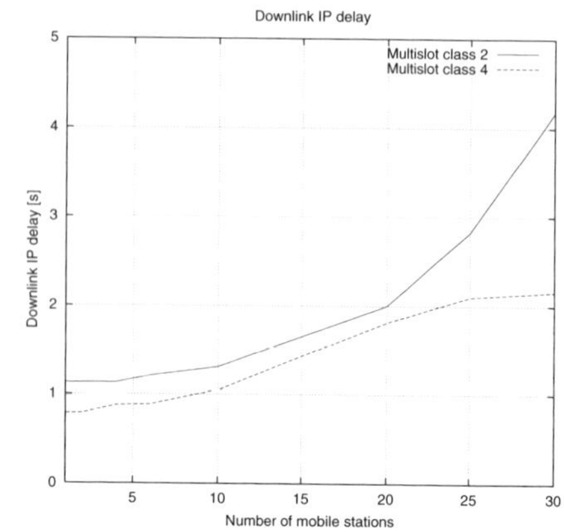

(b) Downlink IP datagram delay

Figure 8.5: Performance measures for different multislot capabilities

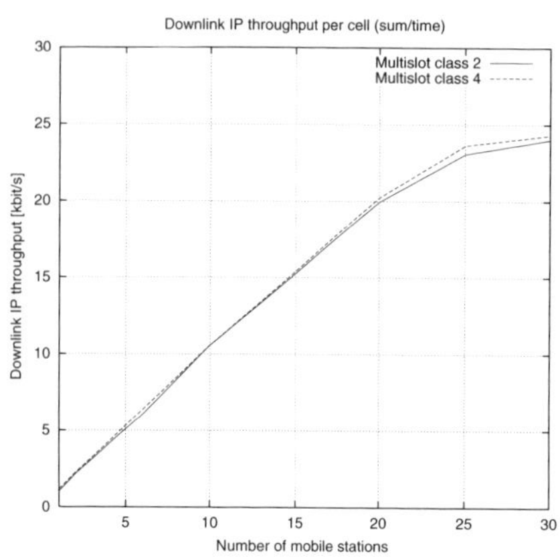

(a) Downlink IP system throughput

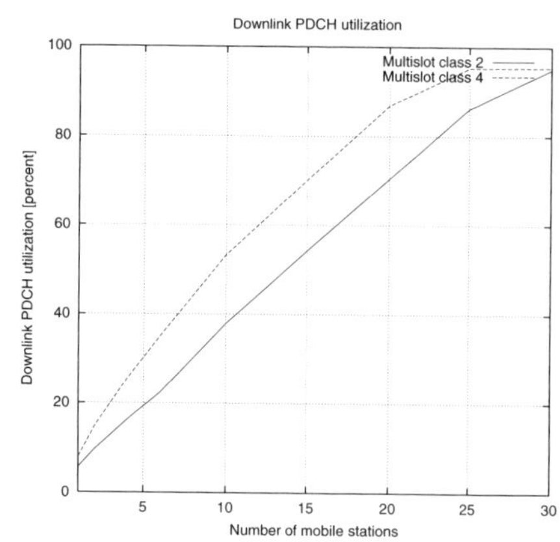

(b) Downlink PDCH utilization

Figure 8.6: System measures for different multislot capabilities

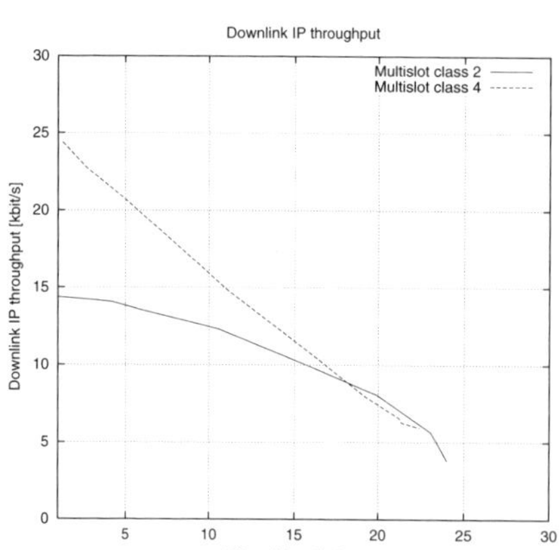

(a) Downlink IP throughput per user

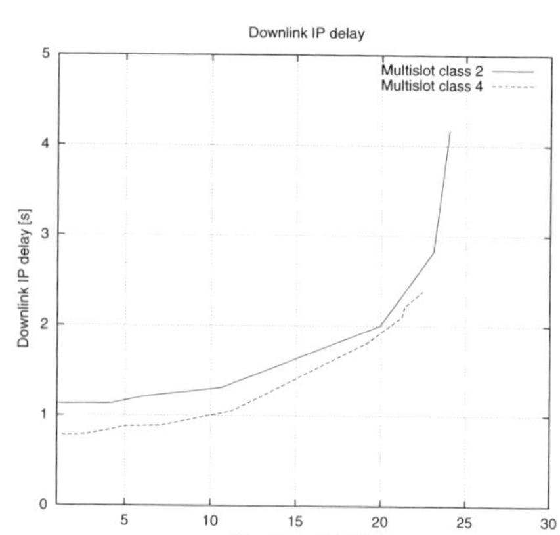

(b) Downlink IP datagram delay

Figure 8.7: Performance measures for different multislot capabilities over offered IP traffic

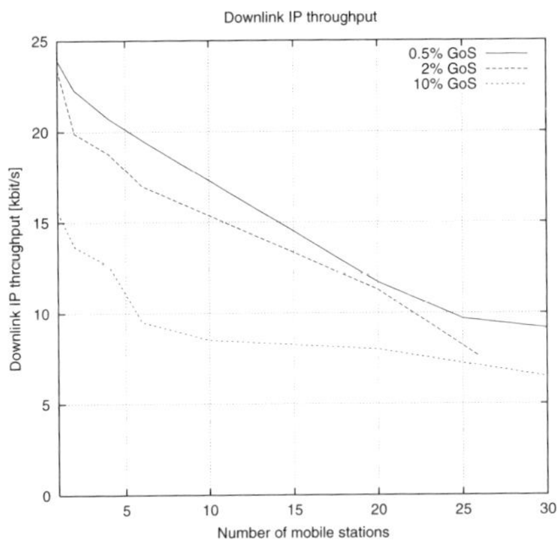

(a) Downlink IP throughput per user

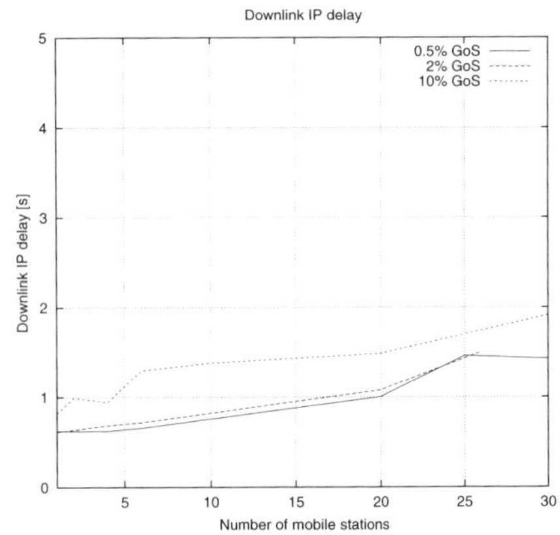

(b) Downlink IP datagram delay

Figure 8.8: Performance measures over the number of MSs

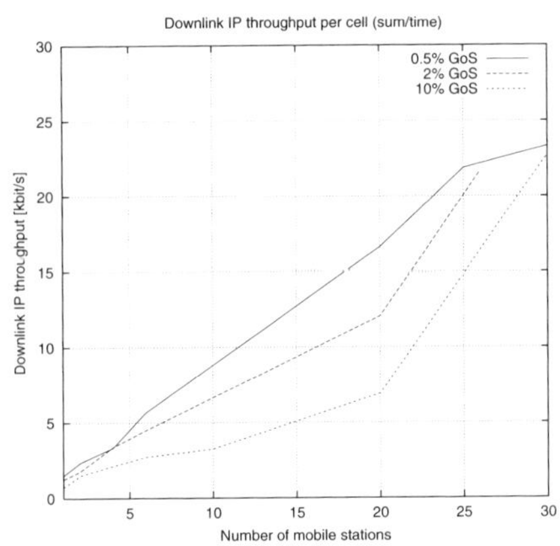

(a) Downlink IP system throughput

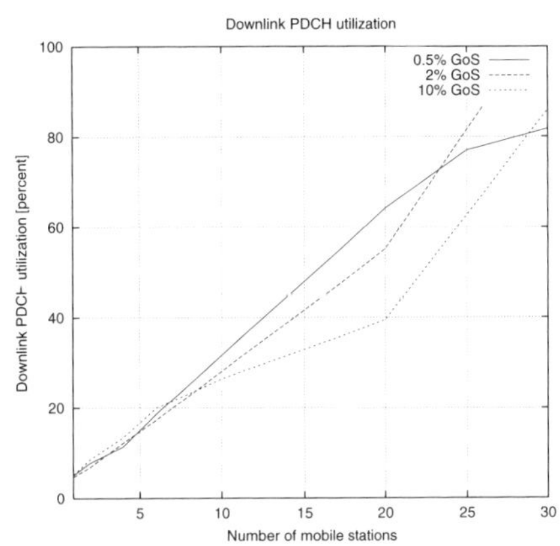

(b) Downlink PDCH utilization

Figure 8.9: System measures over the number of MSs

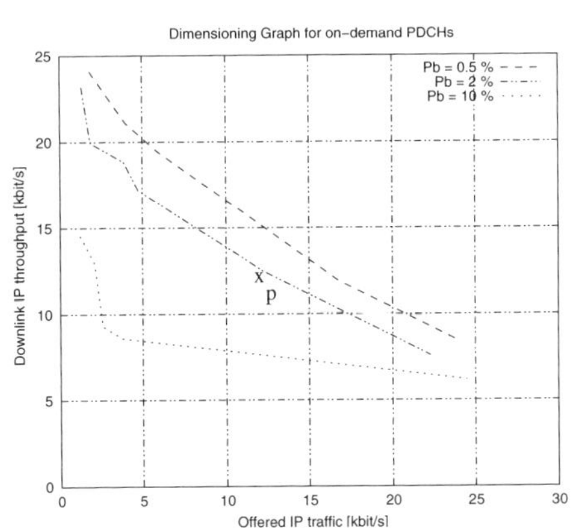

(a) Downlink IP throughput per user

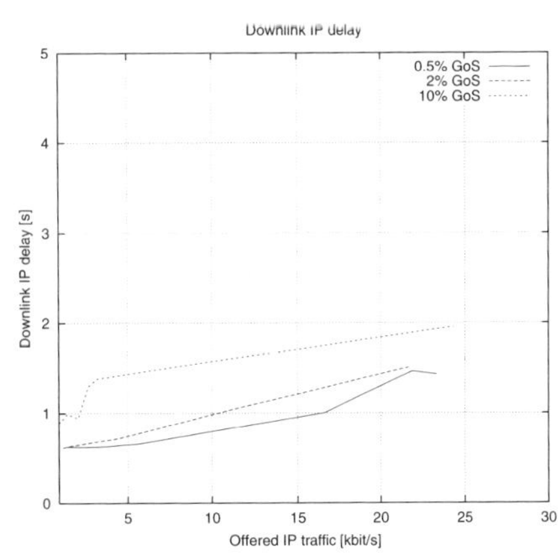

(b) Downlink IP datagram delay

Figure 8.10: Performance measures over the offered traffic

Step 3: Calculate the total offered traffic per cell:

offered traffic/user = 54 kbyte/h = 0.12 kbit/s

total offered traffic = 100 · 0.12 kbit/s = 12 kbit/s

Step 4: Define the desired average user data rate the operator wants to guarantee (here 12.5 kbit/s).

Step 5: Regard the operating point p defined by the desired user performance on the y-axis and the total offered traffic on the x-axis and choose the adequate P_b curve as the next that lies above this operating point p. Here p equals (x = 12 kbit/s, y = 12.5 kbit/s).

If the offered CS traffic corresponding to this P_b is predicted to be exceeded a new TRX should be added. Here p is just below the curve with a P_b of 2%. This means that coexisting CS traffic corresponding to a P_b of 2% is acceptable for this scenario. Since a P_b of 1.5% is assumed an additional TRX would not be needed in this example.

8.3.4 Remarks on Mixed Configurations with Fixed and On-demand PDCHs

To be able to guarantee the availability of GPRS-based services an operator might provide one, two or more fixed PDCHs and might provide the rest (up to eight) as on-demand PDCHs. In this case the results from the pure on-demand configurations can be taken as a worst case estimation, since the probability that the first PDCHs are used by CS traffic is quite low (below 10%) [106].

If four or more fixed PDCHs are provided and the rest (up to eight) are on-demand, there are two possibilities. If an additional TRX is introduced to reduce the blocking probability for CS calls, the on-demand channels will not be highly utilized. Here the capacity can be estimated taking a configuration with eight fixed PDCHs. If no further TRX is provided, the utilization of the on-demand PDCHs will be very high. In this case the capacity can be estimated by a configuration with the allocated number of fixed and no on-demand PDCHs.

8.3.5 Summary of GPRS Performance

For realistic object sizes a maximum IP data rate of about 24 kbit/s was evaluated. Besides the block errors on the radio channel, the protocol overhead of the layers RLC/MAC, LLC and SNDCP and the delay arising with the TBF setup and RLC/MAC acknowledgement messages, the capacity loss arises with the TCP flow control mechanisms.

The performance loss for GPRS if on-demand PDCHs are used depends on the coexisting CS traffic in the regarded radio cell. Since the lower TCH range stays mostly available for GPRS, if the radio network is dimensioned for a P_b of 0.5% and 2%, the performance gain for GPRS achieved by the allocation of fixed PDCHs is only small.

The main cognition is that GPRS remains available with small QoS losses even if there are interceptions enforced by CS connections and the PDCHs are only available on demand. Besides the reason that there is a high probability that a number of channels remains available for GPRS, the flexibility of GPRS has to be mentioned. TBFs can be redistributed on other PDCHs still available. If PDCHs are detracted for a certain period, TBFs can share the resources that are still available and can utilize the resources again when they are reallocated after the CS connection release.

For example, it is shown how the simulation results can be applied for radio network traffic engineering. The dimensioning graph for fixed PDCH configurations is based on simulation results for different load scenarios and profits from the nearly linear relation between downlink IP throughput per user and the offered IP traffic in the radio cell. With this dimensioning graph the relationship between desired QoS, offered traffic and required radio capacity can be estimated.

For on-demand configurations dimensioning graphs based on the simulation results for different CS load scenarios can be used to determine if a new TRX has to be installed to serve the GPRS traffic with a desired QoS.

8.4 Simulation Results for EGPRS and Comparison with GPRS

In this section, the impact of the new *Modulation and Coding Schemes* (MCSs) and the related *Link Quality Controls* (LQCs) functions defined in the EGPRS specifications on the overall system performance and the quality of service (QoS) for the non-real-time Internet applications WWW and e-mail is discussed. The following simulation results are based on the general parameter settings that were introduced in Section 8.1. The cell configuration is given by the number of PDCHs available for GPRS/EGPRS. Here four fixed PDCHs are used. The channel conditions are determined by the cell and cluster size that are the basis for the C/I calculation. Cluster size 3 and 7 and cell sizes with a radius of 300 and 3000 meters are regarded. The same traffic mix and the same protocol settings and capabilities as in Section 8.3 is assumed.

8.4.1 Effect of the Modulation and Coding Schemes

First, the maximum throughput achievable with the EGPRS MCSs is compared with the maximum throughput achievable with GPRS CSs. Additionally the IP delay and the PDCH utilization for the different MCSs are presented. To show the potential in a first step, an error-free channel is assumed.

Figure 8.11(a) shows the mean downlink IP throughput per user as a function of the number of active EGPRS users and the used fixed MCSs. In Figure 8.11(b) the results of the same simulations for GPRS are presented. The EGPRS MCSs 1 to 3 achieve nearly the same performance as the corresponding coding schemes in GPRS. This is plausible since the payload size of the corresponding schemes in GPRS and EGPRS is similar. The performance of MCS-4 differs from that of CS-4. Since the payload size of MCS-4 is smaller than that of CS-4, the throughput of CS-4 is slightly higher.

Under optimal conditions (only one user active, MCS-9, error-free channel) an EGPRS user can reach a maximum mean data rate of 66 kbit/s for the chosen application mix. This is an improvement of about 50% compared with the throughput for CS-4 in GPRS, but starting with 10 users the benefits from EGPRS MCSs become more significant: here the gain rises by over 100%. Figure 8.12 shows the IP packet

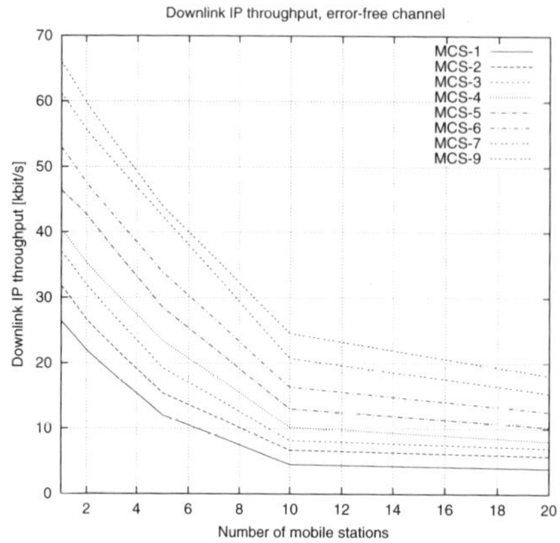

(a) EGPRS

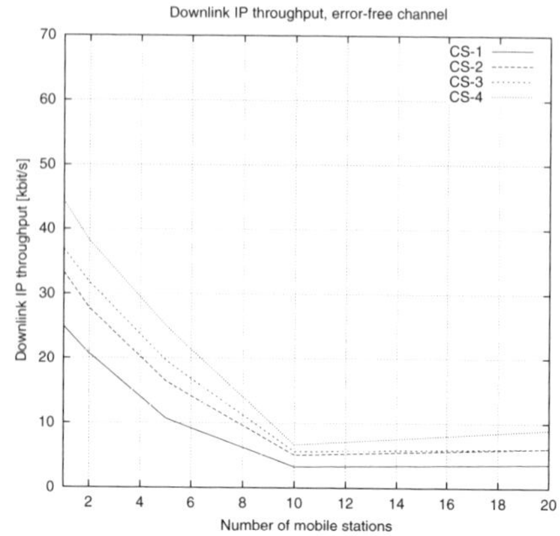

(b) GPRS

Figure 8.11: Mean throughput per user for EGPRS and GPRS (error-free channel)

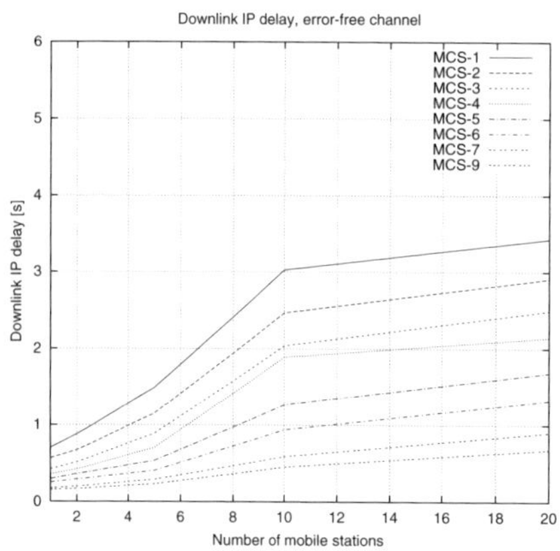

(a) EGPRS

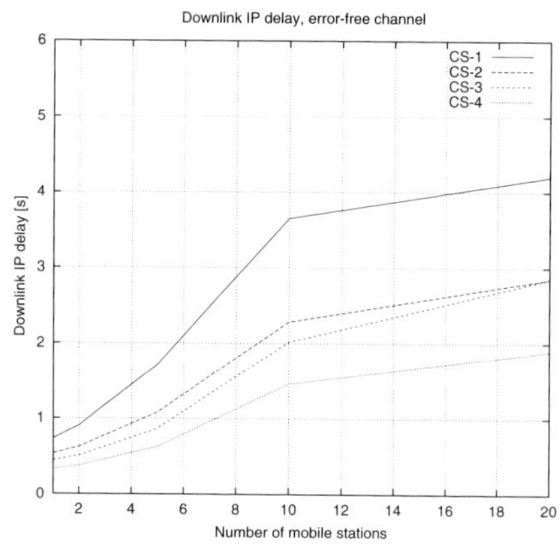

(b) GPRS

Figure 8.12: Mean IP packet delay for EGPRS and GPRS (error-free channel)

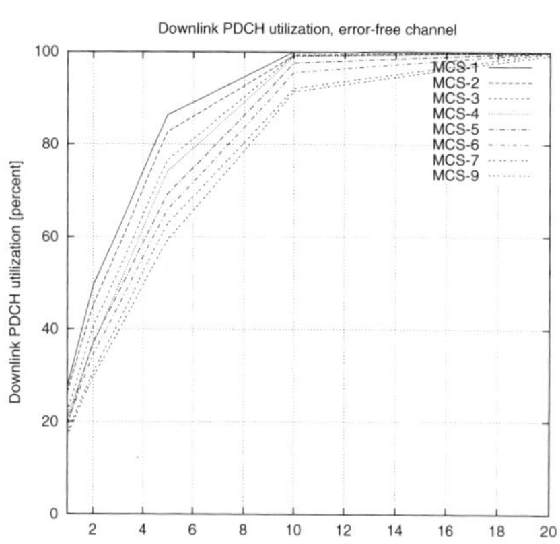

(a) EGPRS

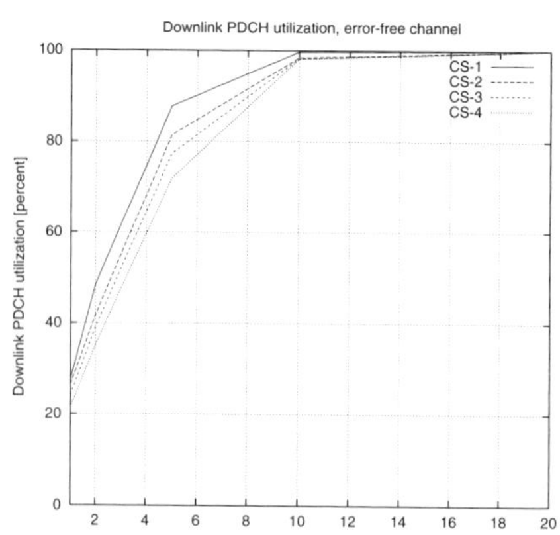

(b) GPRS

Figure 8.13: Mean PDCH utilization for EGPRS and GPRS (error-free channel)

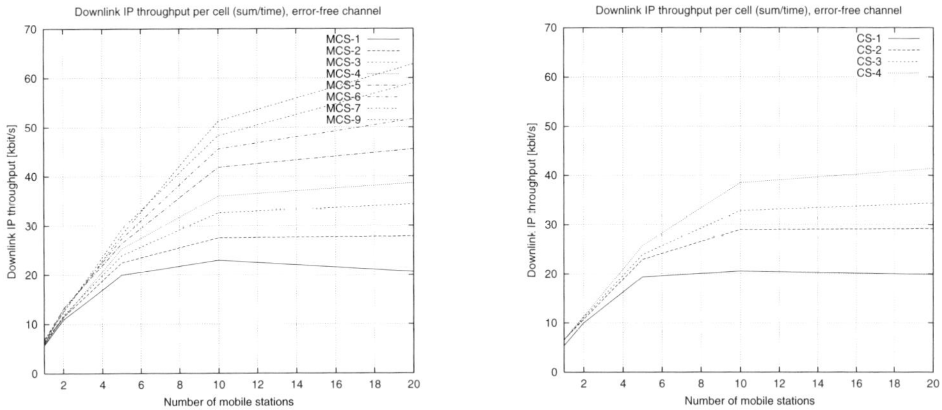

(a) EGPRS (b) GPRS

Figure 8.14: Mean IP throughput in the cell for EGPRS and GPRS (error-free channel)

delay for the different channel coding schemes in EGPRS and GPRS. A comparison shows a significant gain through the EGPRS MCSs.

Next, the system measures are shown. Figure 8.13 compares the PDCH utilization for EGPRS and GPRS; Figure 8.14 compares the mean IP throughput in the cell.

The diagrams in Figure 8.15(a)–(f) show the performance and system measures over the total offered IP traffic. Here the typical curve for a delay is indicated by the pole for MCS-1 and CS-1 and 20 kbit/s offered IP traffic.

The graphical presentation in Figure 8.14 shows that with 10 active MSs the mean utilization of the PDCHs is nearly 100% for all GPRS CSs. The two highest EGPRS MCSs (MCS-7 to MCS-9) at 10 MSs leave some reserves for further users. As expected, the maximum values of the mean cell throughput (Figure 8.15) are reached for the highest MCSs of GPRS and EGPRS (CS-4 for GPRS and MCS-9 for EGPRS). The value for MCS-9 lies between 30% for 10 mobiles up to about 50% for 20 MSs higher than for CS-4.

The results presented within this section do not indicate the maximum throughput, since the evaluated values depend on the application mix and the configuration of, e.g., the WWW sessions.

8.4.2 Effect of LQC on the Block Error Probability

It is expected that LA and IR reduce the error probability of transferred data blocks. Figure 8.16(a)–(d) shows the measured BLEP as a function of the number of MSs and various combinations of the LQC mechanisms. The simulations were carried out for a cluster size of 3 and 7. Since the mean channel quality within a cell with cluster size 3 is lower than within a cell with cluster size 7 the positive effect of IR is more significant. Also the differences of LA and init LA regarding the BLEP become more significant. Init LA means that the MCS only can be switched from the old to the new TBF. For cluster size 3 two velocities, 6 and 100 km/h, were examined.

The performance measures such as throughput and delay cannot be directly derived from the BLEP values. If, e.g., all data blocks are coded with MCS-1 the error rate

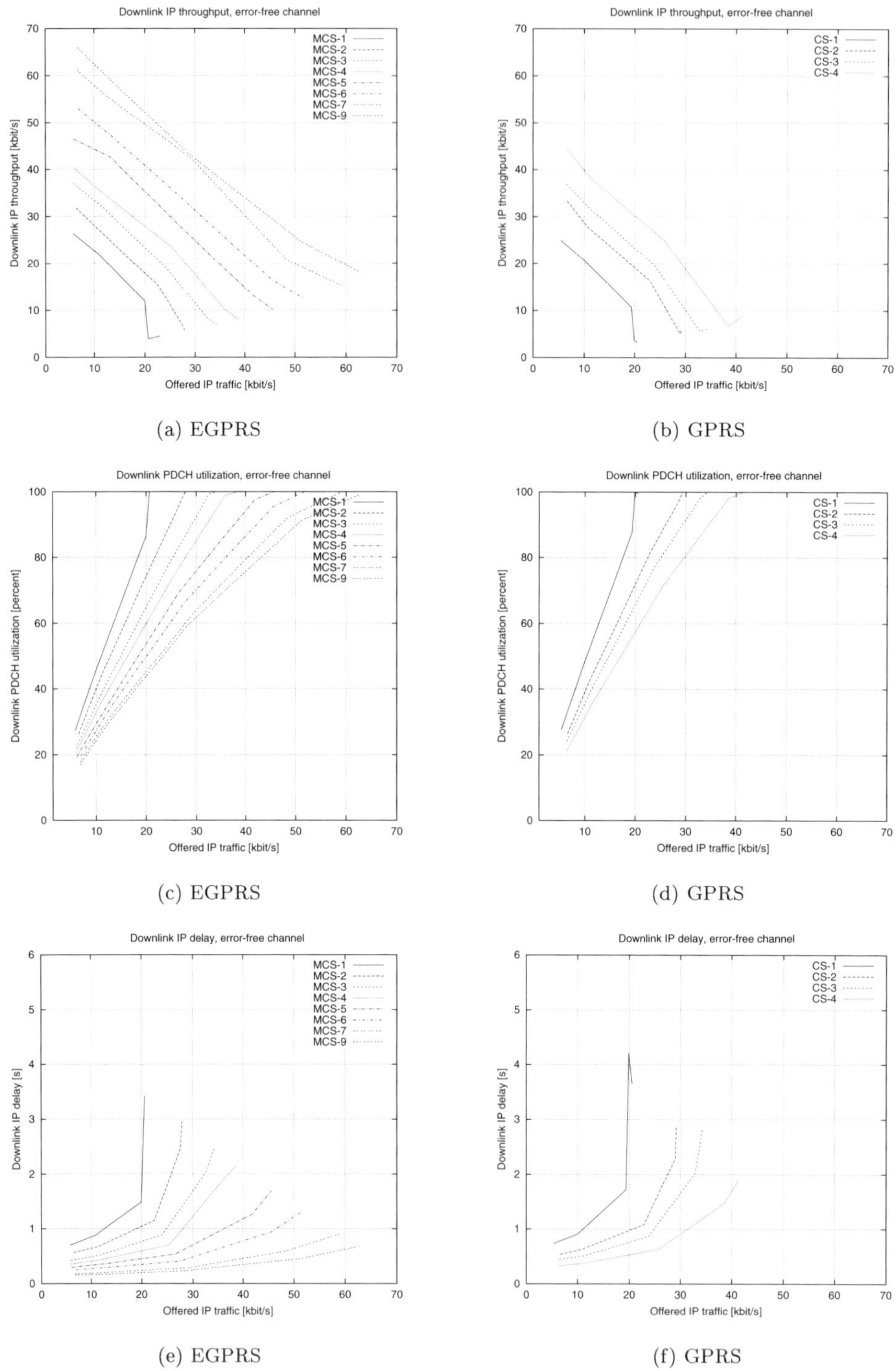

Figure 8.15: Mean IP throughput, PDCH utilization and IP packet delay versus offered traffic for EGPRS and GPRS (error-free channel)

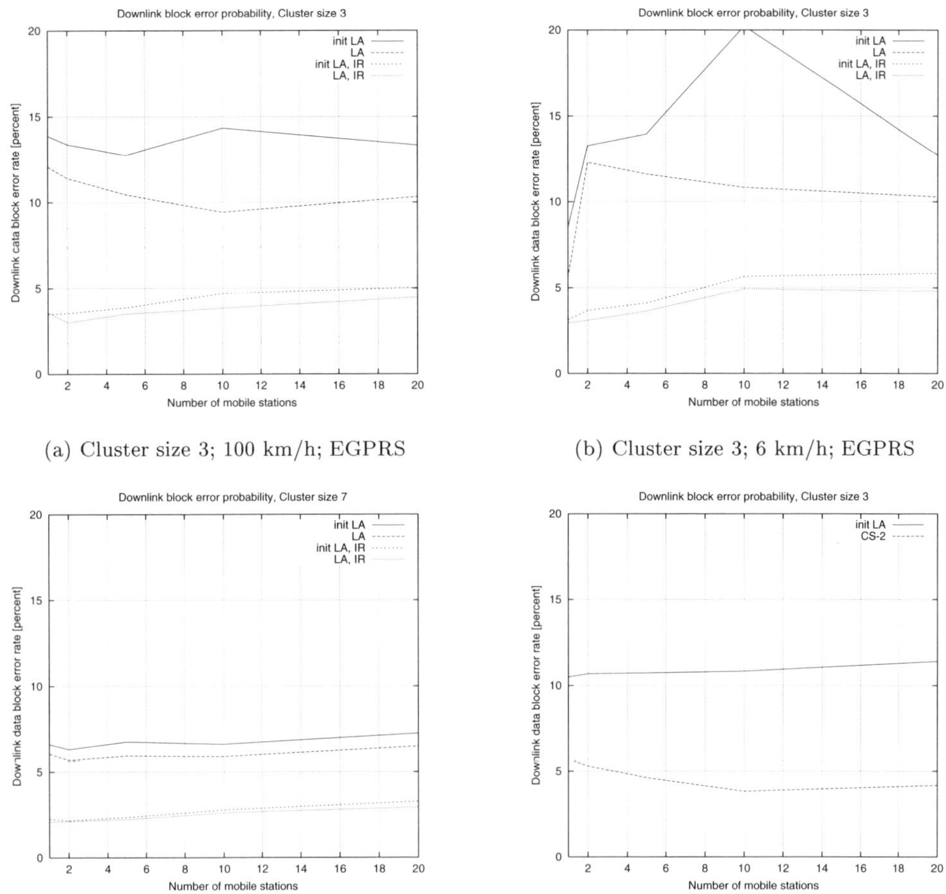

(a) Cluster size 3; 100 km/h; EGPRS

(b) Cluster size 3; 6 km/h; EGPRS

(c) Cluster size 7; 100 km/h; EGPRS

(d) Cluster size 3; 100 km/h; GPRS

Figure 8.16: Downlink data block error probabilities (BLEP)

is very low, but since the payload of MCS-1 coded blocks is very small, only low data rates can be reached.

8.4.3 Performance Evaluation Regarding Different LQC Mechanisms

Now the performance and system measures, especially the IP throughput performance, are discussed considering the EGPRS LQC functions LA and IR compared to fixed MCSs.

First, a scenario with cluster size 3, cell radius 300 m and MS velocity 6 km/h is regarded. Simulations have shown that fixed MCS-5 reaches the best performance for this scenario compared to other fixed MCSs.

Additionally, the performance is considered for LA without IR and LA with IR. The latter represents full LQC capability.

Figure 8.17(a) shows that the throughput performance with LA is increased by up to 20% compared to the fixed MCS-5. If additionally IR is used, the performance gain is not significant to LA without IR. In Figure 8.17(b) the mean downlink IP packet delay is presented. Compared to the operation with the well-chosen fixed MCS-5, IR is able to decrease the delay by up to 10%. LA achieves a delay performance gain of even 20 to 25%, while the additional use of IR does not change the delay performance significantly.

The LQC functions do not have a major effect on the system throughput per cell and the PDCH utilization compared to the well-chosen fixed MCS-5 (see Figure 8.17(c) and Figure 8.17(d)). In the range of 10 to 20 MSs the cell throughput ranges between 37 and 41 kbit/s. In Figure 8.17(e) and Figure 8.17(f) the performance measures throughput and delay are shown over the offered IP traffic.

Now the throughput and delay performance is discussed regarding the LQC functions in comparison to a less robust fixed MCS chosen. If LA is not used and a less robust MCS is chosen, e.g. MCS-6, the throughput performance can be increased with IR. Figure 8.18(a) shows a gain of up to 30%. If the MCS is chosen even more aggressively, e.g. MCS-9 (see Figure 8.18(c)), the performance cannot be fully recovered compared to fixed MCS-5.

Finally, the throughput performance considering different cluster and cell sizes is considered. Figure 8.18(e) shows that the mean downlink IP throughput per user during transmission periods reaches up to 45 kbit/s in the scenario with cluster size 7 and 3000 m cells.

While the performance gain in this scenario compared to the scenario regarded above ranges between 15% and 30% in low load situations, an increase of up to 45% is reached in high load situations. The throughput does not fall below 14 kbit/s for the scenario with cluster size 7 even with 20 MSs that are active in the radio cell.

8.4.4 Summary of EGPRS Performance

The performance and system measures of EGPRS regarding non-real-time Internet applications were presented. Different LQC functions and their impact on the overall performance are examined.

For the regarded traffic mix (30% WWW, 70% e-mail), a throughput performance gain of up to 100% compared to GPRS is reachable with the new MCSs in EGPRS. A maximum downlink IP data throughput per user during transmission periods of about 66 kbit/s is reachable for this traffic mix with four PDCHs available and the multislot capability of four slots in up- and downlink.

Additionally LA and IR are enabling performance gains in comparison to a fixed chosen MCS. Although the gain through LA compared to a well chosen fixed MCS is not dramatic, LA should be used in EGPRS networks. The reason is that without LA, a fixed MCS would have to be chosen for every site. This cannot be simply realized in realistic radio coverage planning processes.

8.5 Performance of Evolving Mobile Applications

This section presents traffic performance results for different traffic mixes. These results are useful to estimate the radio capacity that is needed during the evolution of GSM networks towards third-generation (3G) mobile communication systems. First

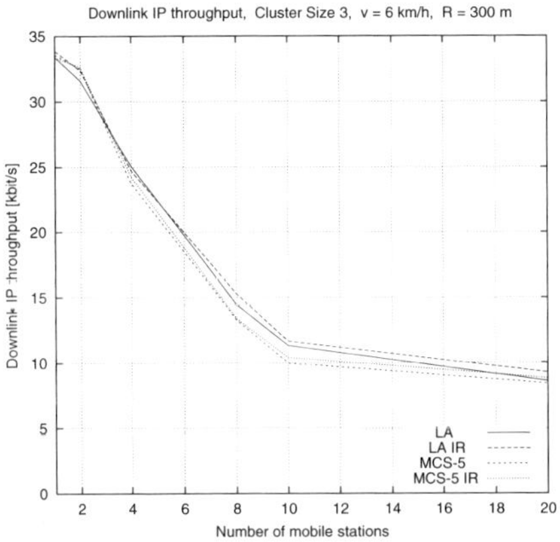

(a) Mean DL IP throughput per user

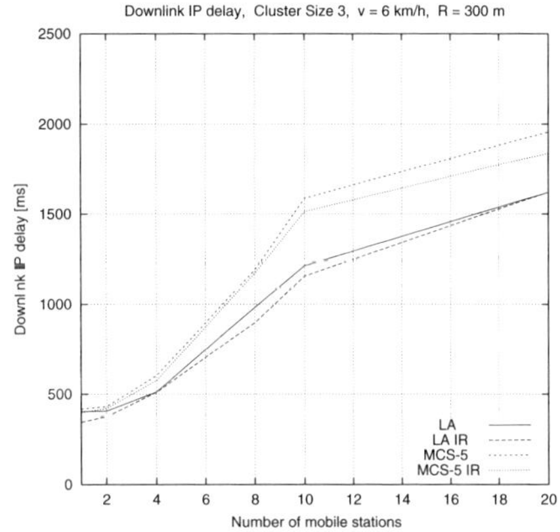

(b) Mean DL IP packet delay

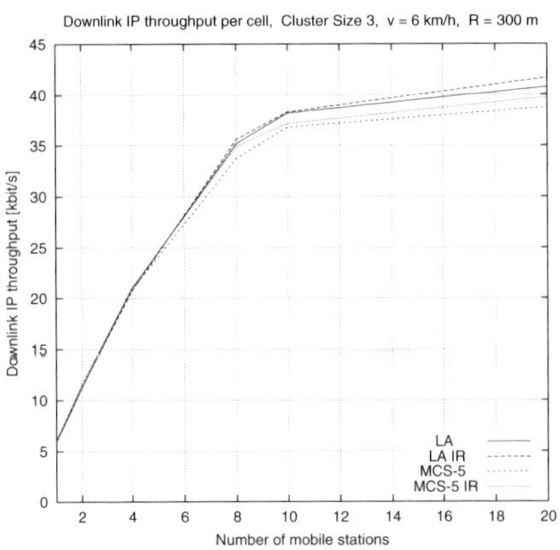

(c) Mean DL IP throughput per cell

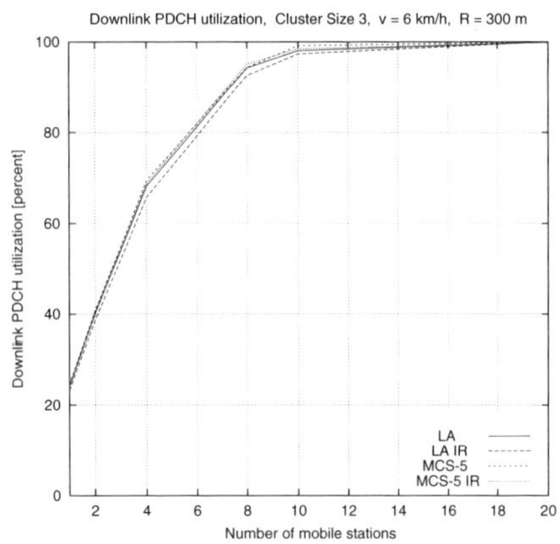

(d) Mean DL PDCH utilization

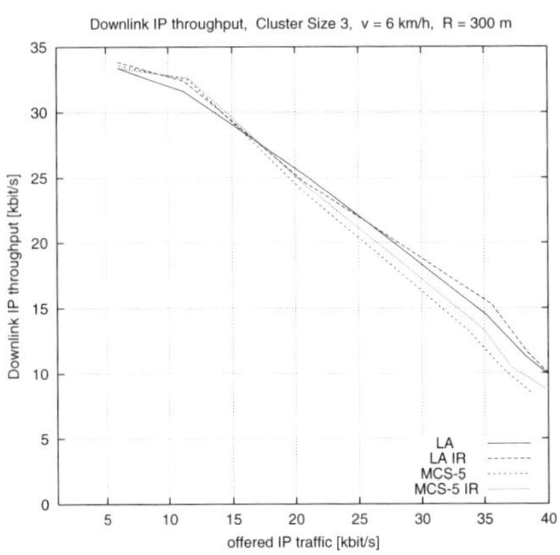

(e) Mean DL IP throughput per user

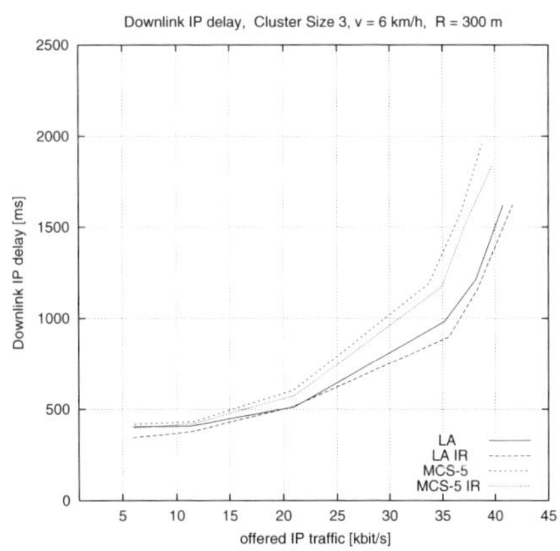

(f) Mean DL IP packet delay

Figure 8.17: Performance and system measures for different LQC capabilities

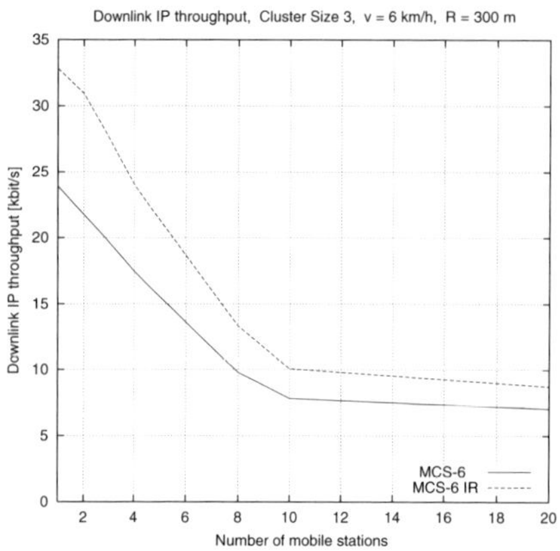

(a) Mean DL IP throughput per user

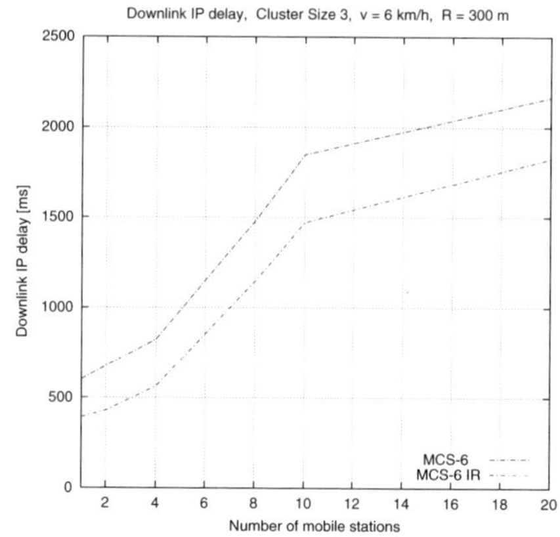

(b) Mean DL IP packet delay

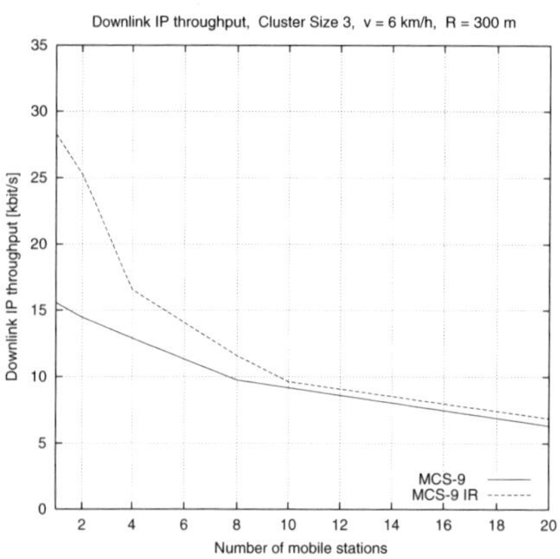

(c) Mean DL IP throughput per user

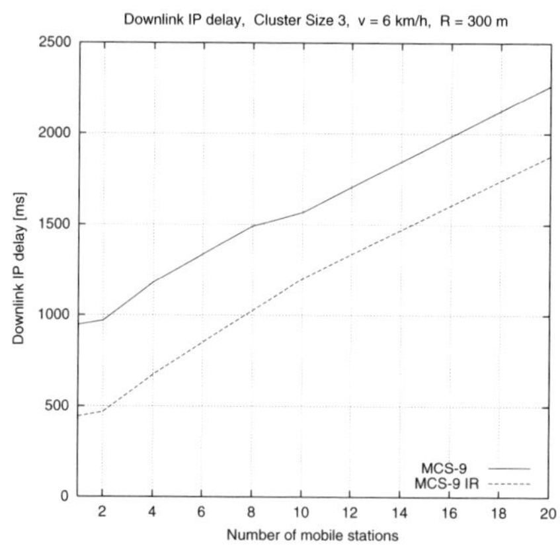

(d) Mean DL IP packet delay

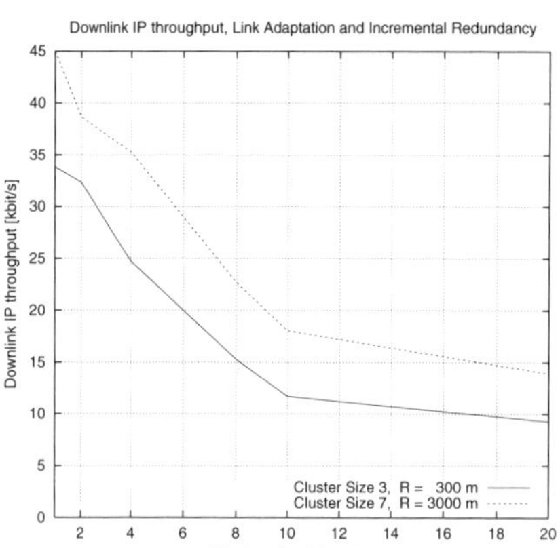

(e) Mean DL IP throughput per user

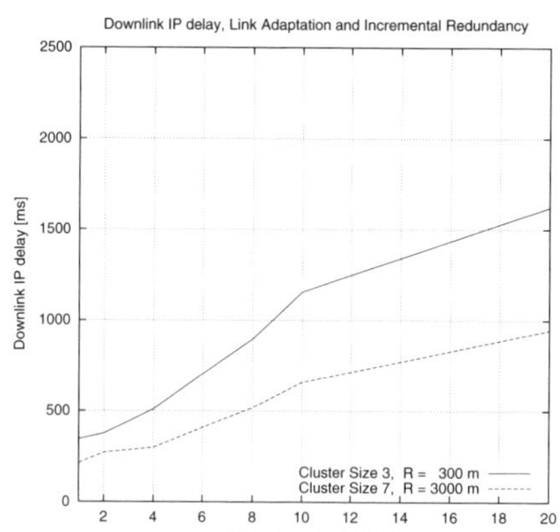

(f) Mean IP packet delay

Figure 8.18: Performance measures for different fixed MSCs and IR a traffic mix of applications based on the Wireless Application Protocol (WAP) and conventional Internet applications like WWW browsing and e-mail over the General Packet Radio Service (GPRS) are regarded. In the next step traffic performance results for Streaming applications over GPRS and EGPRS are presented and the feasibility of Streaming with coexisting interactive and background applications like WWW and e-mail is regarded. The traffic characteristics of these applications are represented by the traffic models proposed in Chapter 6.

8.5.1 WAP in Comparison to Internet Applications over GPRS

To be able to compare the user-perceived performance of WAP in comparison to conventional Internet applications, the application response time is shown in Figure 8.19(a) for pure WWW, e-mail and WAP traffic.

In situations with low traffic load the response time for a WAP deck is below 2 s, while the response time for a web page is around 4 s. The reason is that a web page has a larger content size and is transmitted over TCP.

In load situations with higher traffic load the response time for a WAP deck remains nearly constant for up to 20 MSs. If only one PDCH is available, the WAP response time increases to more than 10 s for 20 MSs in the radio cell. Because of the larger content size the response time for web pages exceeds 20 s already with 10 active MSs in the radio cell even if four PDCHs are available. The reason for the strong increase in response time for WWW and e-mail can be seen in other evaluated measures like the downlink PDCH utilization in Figure 8.21(a). More than 95% PDCH utilization is reached for WWW/e-mail traffic with 15 MSs, while 15 WAP users are utilizing the PDCHs with only 30% for the same PDCH configuration.

Figure 8.20(a) shows the mean downlink IP throughput per user during transmission periods. While the throughput performance for pure WAP traffic remains relatively constant with an increasing number of mobile stations and four PDCHs, it decreases dramatically for pure WWW/e-mail traffic because of the higher offered traffic and the higher utilization. The poor throughput performance for WAP traffic can be explained by the low WAP deck size. Such transaction-oriented applications are more influenced by the high round-trip time, which is mainly caused by the high delay over the air interface, than by the available bit rate. Since the response time for a WAP deck is less than 1.5 s, which should be acceptable for a wireless application, the user is not aware of this low throughput performance.

Since WWW and e-mail applications comprise larger file sizes to download than WAP-based applications do, the throughput performance perceived by a user in situations with low traffic load ranges from 14–24 kbit/s. These performance values are mainly influenced by the characteristics of the offered traffic. Since the e-mail traffic model has larger file sizes than WWW, the throughput performance is better. With the number of mobile stations increasing up to 15, saturation is reached and the performance for WWW and e-mail users gets unacceptable and even gets worse than the low throughput for pure WAP traffic. In this situation with high traffic load the WWW and e-mail traffic performance is less influenced by the characteristics of the traffic model such as file size, and more by the load on the air interface.

8.5.2 Traffic Mix with WAP and WWW/e-mail over GPRS

Since the predicted traffic mix for GPRS networks will be composed of WAP traffic and conventional Internet applications like WWW and e-mail, the GPRS traffic performance for a traffic mix of 60% WAP, 28% e-mail and 12% WWW sessions will be regarded here.

Figure 8.19(b) shows the application response time for WAP decks, e-mails and WWW pages, respectively. Compared to the graphs in the previous section, the WWW and e-mail performance is not strongly affected by WAP traffic, since small WAP packets can be multiplexed seamlessly with the TCP-based WWW and e-mail traffic. The throughput (see Figure 8.20(b)) decreases more slowly with an increasing number of mobile stations than in Figure 8.20(a) with pure WWW, e-mail and WAP traffic regarded separately, since here WAP represents the main part of a traffic mix and the total offered traffic per radio cell is increasing much more slowly.

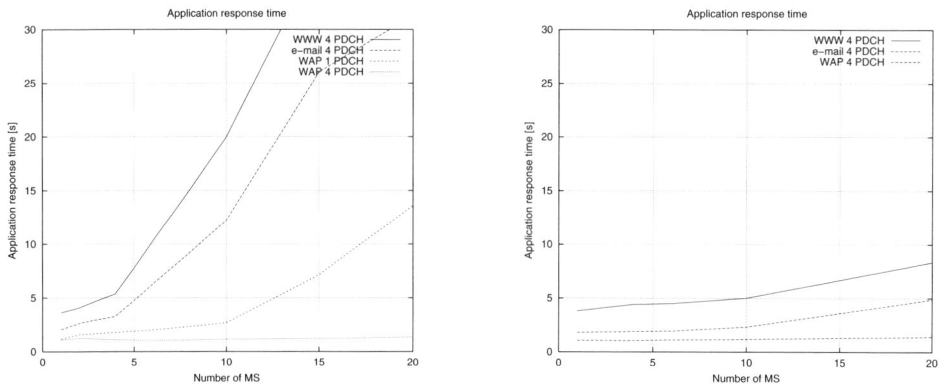

(a) Pure WWW/e-mail and WAP traffic (b) Traffic mix

Figure 8.19: Mean application response time

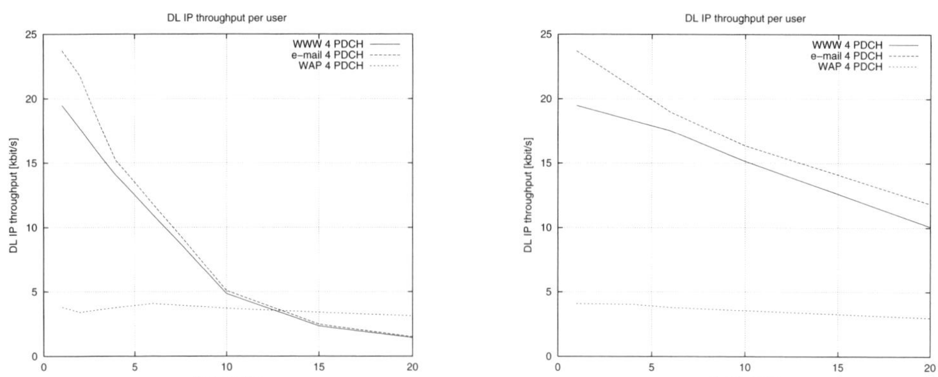

(a) Pure WWW/e-mail and WAP traffic (b) Traffic mix

Figure 8.20: Mean downlink IP throughput per user

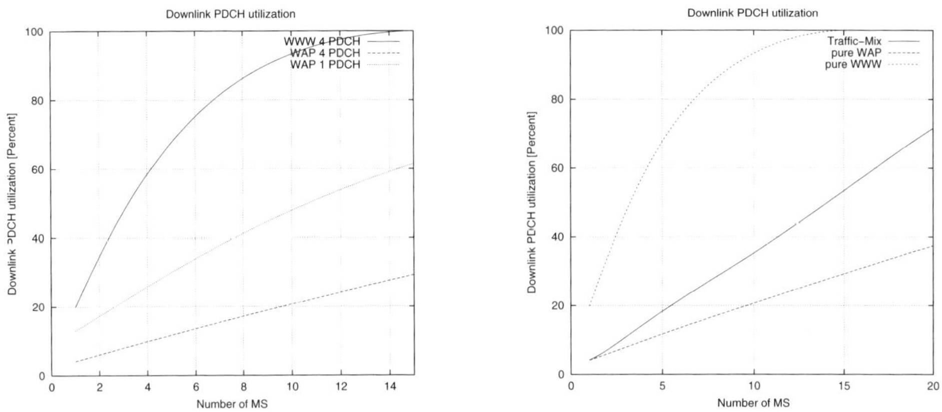

(a) Pure WWW/e-mail and WAP traffic

(b) Traffic Mix compared to pure WWW and pure WAP traffic

Figure 8.21: Mean downlink PDCH utilization

The same applies for the response time. In the scenario with traffic mix WWW pages have a response time of 5 s with 10 active stations generating a traffic mix, while 10 stations generating pure WWW traffic have to wait for more than 20 s.

The WAP response time increases slightly from 1.2 s for pure WAP traffic to 2.1 s for the traffic mix scenario. The reason is that WWW and e-mail sessions are composed of larger application packets that leave less resources open for WAP users (see Figure 8.21(b)). Nevertheless a response time for WAP decks of 2.1 s still should be acceptable.

8.5.3 Video Streaming over GPRS and EGPRS

As a typical EGPRS introduction scenario, Video Streaming applications over EGPRS are examined in coexistence with WWW and e-mail applications. Due to the conservative predictions concerning the future usage of Streaming applications the mix contains only 10% Streaming sessions. The remaining part is assumed to 63% e-mail and 27% WWW session.

Cluster size 7, a cell size with a radius of 3000 meters and a velocity of 6 km/h have been regarded. Both Link Adaptation (LA) and Incremental Redundancy (IR) are applied.

As the critical performance measure the downlink IP datagram delay is regarded (see Figure 8.22(a)). Additionally the downlink IP throughput per user indicates if the Streaming data rate of 14.39 kbit/s can be maintained under a regarded number of mobile stations offering the Multimedia traffic mix (see Figure 8.22(b)). In addition the distribution of the downlink IP datagram delay for one and 10 mobile stations is shown in Figure 8.23(a) and Figure 8.23(b).

With four available PDCHs in the regarded radio cell the IP datagram delay for Streaming increases dramatically to more than 10 s with more than six active stations that generate a traffic mix. With more than six stations the throughput of 14.39 kbit/s (see Section 6.5.1.3) can not be maintained, which shows that the performance for

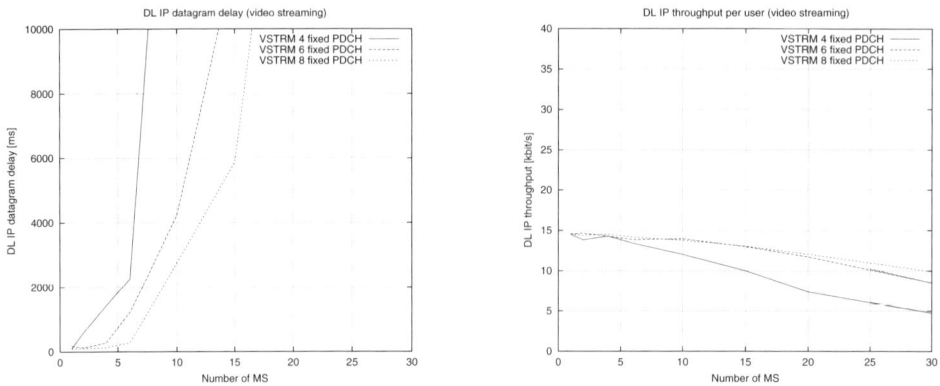

(a) Mean downlink IP datagram delay (b) Mean downlink IP throughput per user

Figure 8.22: Performance of Video Streaming applications (traffic mix)

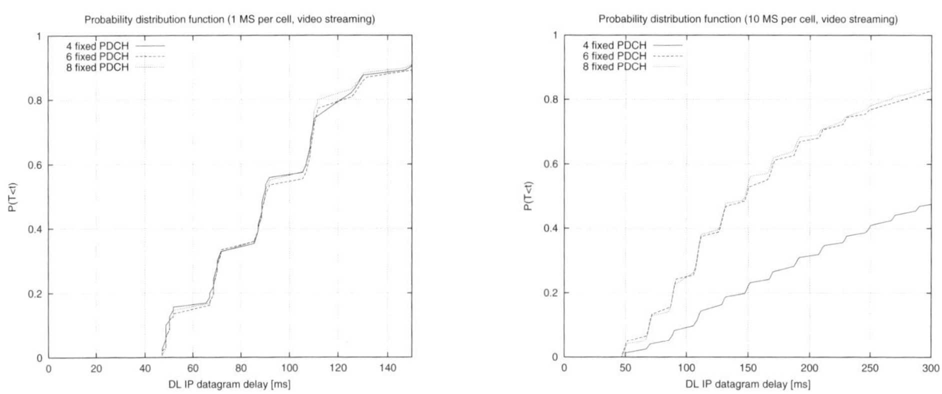

(a) One mobile station (b) 10 mobile stations

Figure 8.23: Downlink IP datagram delay distribution for Video Streaming applications (traffic mix)

Streaming users becomes unacceptable. With six and eight PDCHs available in the cell, 15–20 users generating the traffic with 10% Streaming can be satisfied. The delay starts increasing dramatically with 15 and 20 users, respectively.

Regarding the downlink IP throughput per user in Figure 8.22(b) there is no significant difference in the performance between six and eight available PDCHs. The throughput for four, six and eight PDCHs starts at the same level of 14.39 kbit/s. This is exactly the data rate needed for the chosen video sequence. The downlink IP throughput per user remains constant as long as the necessary data rate for Streaming is provided. Depending on the number of fixed PDCHs the real-time data rate decreases below the required rate of 14.39 kbit/s. At this point the IP datagram delay increases dramatically. With 15 users the required data rate can no longer be maintained.

The distribution functions in Figure 8.23(a) and Figure 8.23(b) confirm this interpretation. For one mobile station 90% of the IP datagrams for the Streaming applications are delivered within 150 ms. The performance does not depend on the number of PDCHs available, since the regarded mobile stations can use only a maximum four slots on the downlink. For 10 mobile stations and four PDCHs available more than 50% of the IP datagrams are delayed more than 300 ms and the slow increase of the distribution function indicates a high delay variance, which makes the delay performance for Streaming applications unacceptable. With six and eight PDCHs 85% of the IP datagrams for Streaming applications are delivered within less than 300 ms, which makes the Streaming performance just acceptable for 10 users generating the Multimedia traffic mix. The steps in the distribution functions are affected by the segmentation of IP datagrams into radio blocks and the number of radio blocks transmitted within a GPRS radio block period of 20 ms.

The different requirements of the applications can be supported by Quality of Service (QoS) management functions in the RLC/MAC layer (see [104, 107]). The transmission of Streaming data may be privileged at the expense of background traffic. While the application response times for WWW and e-mail would increase, the Streaming application would be able to proceed, although high background traffic load would occur in the cell.

8.5.4 Summary of the Traffic Performance of Evolving Applications

The performance of different Multimedia applications in packet-switched cellular radio networks based on GPRS and EGPRS was presented. For GPRS introduction and evolution scenarios WAP applications and a traffic mix of WAP and conventional Internet applications over GPRS are examined. After the performance characteristics of WAP and Internet applications have been regarded separately, the effects of coexisting Internet traffic on WAP traffic and vice versa are outlined. It has been shown that WAP traffic can be multiplexed seamlessly with the Internet traffic because of the small and limited WAP deck size, while Internet traffic slightly slows down WAP traffic in situations with high traffic load. Regarding Video Streaming applications in coexistence with TCP-based applications over EGPRS it has been shown that only a small number of Streaming users can be served by EGPRS, even if the percentage of Streaming in the traffic mix is low. At least four fixed PDCHs should be available to support Streaming applications together with background TCP traffic. Privileged transmission of real-time data, realized by QoS management, is one approach to providing the required bit rate for video Streaming in situations with high traffic load.

8.6 Performance Gain through QoS Management

While in the first phase after GPRS introduction only best-effort data services without differentiating subscribers and applications will be supported, in the second phase quality of service (QoS) management functions will be integrated to be able to guarantee subscriber- and application-specific QoS requirements (see Section 4.4.2 and Section 7.5). These QoS functions will be based on the aggregation of flows belonging to the same service class and the prioritized admission control and scheduling of these aggregate flows in the radio network. No strict resource reservation is performed. This is the motivation to use DiffServ in GPRS/EDGE core networks [83, 44], since DiffServ is also a scalable approach without need for per-flow state information and signaling at every hop, in contrast to the Integrated Services (IntServ) [40] approach, which is based on the Resource Reservation Protocol (RSVP) [39].

For radio network dimensioning and network equipment further development the effect of these QoS management functions on the overall system performance has to be determined. In this context we do not aim at optimized solutions for QoS management functions, but at an estimation of the performance gain achieved with their introduction compared to a pure best-effort service.

First the performance gain through QoS functions in the radio network is examined compared to a pure best-effort service. Since radio resources are scarce and represent the system bottleneck it is firstly assumed that the core network is well dimensioned. In the next step an IP-based core network with and without DiffServ capabilities is regarded. The feasibility of QoS support in mobile core networks interworking with the QoS functions in the radio network is examined for low traffic load situations and under congestion.

8.6.1 Scenario Parameter Settings

Here a typical three-TRX scenario is regarded with zero and one fixed and eight and seven on-demand PDCH that are shared with circuit switched traffic, which is offered corresponding to an Erlang-blocking probability of 1% (see Section 8.3.3). This means that on average around seven PDCHs are available for GPRS [106].

The Internet traffic is composed of 70% e-mail sessions and 30% WWW sessions (see Chapter 6) not depending on the subscription profile of the regarded MS. The traffic classes Streaming and Conversational are not regarded here, since the first GPRS usage will be dominated by browsing and messaging applications. 10% of the mobile stations are representing Premium subscribers and 90% Standard subscribers. The values of the CAC limits P and I (see Section 7.5.4) are set to one flow per available PDCH for the Premium class and 10 flows per available PDCH for the Interactive class.

8.6.2 Simulation Results Neglecting the Core Network

Figure 8.24(a) shows the mean downlink IP system throughput per radio cell for no and one fixed PDCH and with and without QoS management functions. The difference between the curves with no and one fixed PDCH is very small since only for 1% of the time are all PDCHs allocated for circuit-switched calls. Since the offered circuit-switched traffic is lower for the one-fixed-PDCH scenario, the system throughput is 1–4% higher in the no-fixed-PDCH scenario. As expected the system throughput for low load situations with fewer than 20 MSs in the cell is nearly the same for the results for a best effort (BE) service and a service with QoS functions. In higher load situations the system throughput becomes saturated. This can be explained by the effect that up to 28% of the Background sessions are terminated when no IP datagrams are received for a period of more than 30 seconds (see Figure 8.26(a)). This does not occur in the BE simulations. The same effect can be seen in Figure 8.24(b), where the channel is not utilized with more than 75% in the results with QoS functions.

In Figure 8.25(a) the downlink IP throughput per user during transmission periods for the different service and subscriber classes Premium, Interactive and Background compared with simulation results for a pure BE service (without CAC) is presented. In situations with low traffic load Standard users are losing 15–20% of performance

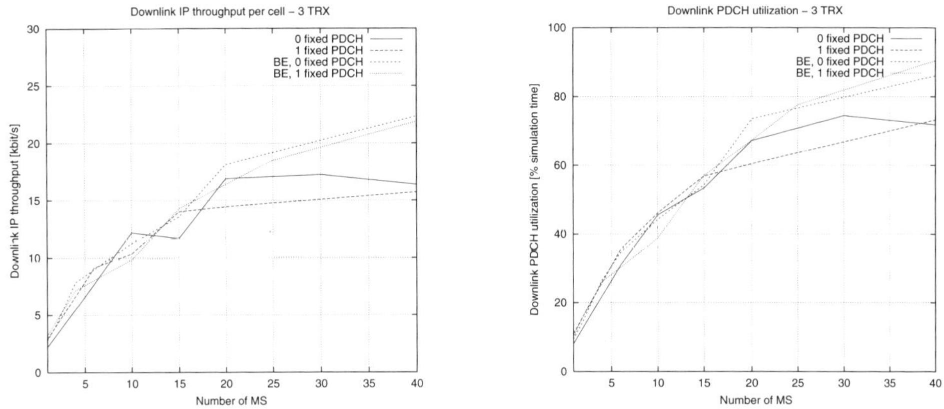

(a) Mean downlink IP system throughput per cell

(b) Mean downlink PDCH utilization

Figure 8.24: System measures with and without QoS management functions

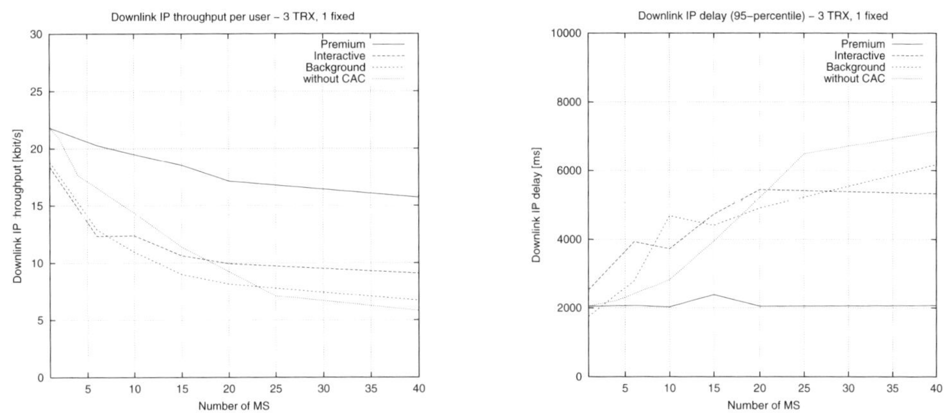

(a) Mean downlink IP throughput per user

(b) 95-percentile of downlink IP packet delay

Figure 8.25: Performance measures for different subscriber and application classes

compared to the BE service while the Premium user performance always remains higher than 15 kbit/s. In higher load situations the service differentiation between Interactive and Background becomes visible. While the throughput performance of the Interactive traffic does not fall below 10 kbit/s, the performance loss for Background applications is not visible in this measure. Nevertheless, more than 20% of the Background sessions are terminated because of poor performance as mentioned above. This can be avoided using fairer scheduling algorithms. The 95-percentile of the IP packet delay in Figure 8.25(b) shows a similar effect.

8.6.3 Simulation Results Considering the Core Network

The core network model is configured with an output rate for each router of 2 Mbit/s. Although core IP routers might have higher output rates, the effect of backbone traffic on the GPRS traffic performance can be modeled with this configuration. Scheduling periods comprise 40 ms. Additional backbone traffic of 1 Mbit/s and 2 Mbit/s is generated to model a low load scenario and a congestion scenario in the core network, respectively.

For each background load scenario, one simulation series is performed with QoS management in the radio and core network (interactive and background) and one with a pure best-effort service without service differentiation in the radio and core network.

8.6.3.1 Low Backbone Traffic Load

Scenarios were examined for service differentiation in the radio and core network. They are compared with the performance of the same traffic mix of WWW and e-mail, if no service differentiation in the radio and core network is done. Since the resources in the core network are not highly utilized by backbone traffic the core network can be seen as nearly transparent even for the best-effort GPRS traffic.

Results very similar to the results shown in Figure 8.25(a) and Figure 8.25(b) were measured. In this scenario with 2 Mbit/s router output rate and only 1 Mbit/s backbone traffic DiffServ does not have a significant influence on the performance, since there is enough capacity left for the regarded GPRS traffic. However, it can be stated that the prioritization of the Interactive class is supported by DiffServ and the results are nearly equal to the case where the influence of the core network is neglected.

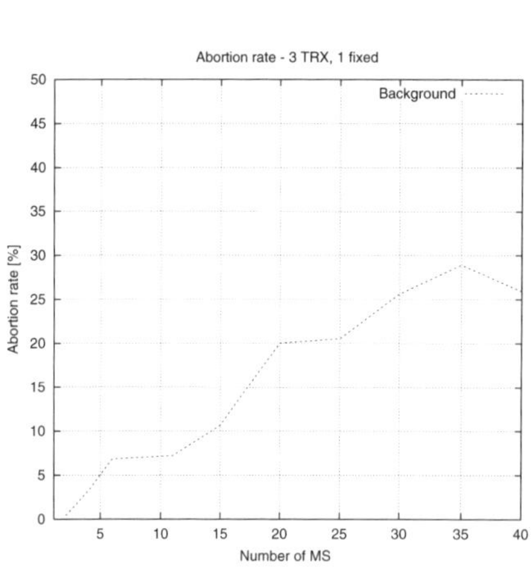

(a) Abortion rate for Background applications

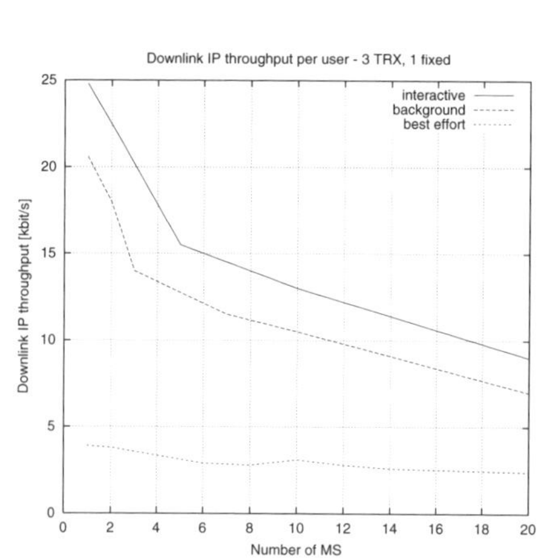

(b) Downlink IP throughput per user

Figure 8.26: Performance measures for high backbone traffic load

8.6.3.2 Congested Core Network

More significant effects of DiffServ on the performance of different service classes become visible in a scenario with high backbone traffic load. In the regarded scenario the backbone traffic equals the output rate of the routers in the core network. Since the best-effort GPRS traffic will be served with the same priority as the backbone traffic, the congestion in the core network will affect the GPRS traffic. This results in poor performance for best-effort GPRS traffic. This is shown in Figure 8.26(b) with average IP throughput values below 5 kbit/s even in situations with low GPRS traffic load. If WWW and e-mail traffic is not served as best-effort traffic, but service differentiation in both the radio and the core network is performed, the same performance as in a low utilized core network can be reached through DiffServ QoS functions. Figure 8.26(b) shows that average downlink IP throughput values above 20 kbit/s are achieved in situations with low GPRS traffic load. With increasing GPRS traffic the performance decreases similarly as in the scenario without core network congestion (see Figure 8.25(a)). This is only caused by the limited resources of eight on-demand PDCHs in the GPRS radio network.

8.6.4 Summary of the Performance Gain through QoS Management

The capacity and performance gain achievable with quality of service functions in GPRS networks comprising Connection Admission Control (CAC) and scheduling with subscriber and service differentiation is examined. Simulation results show that Premium users can be served with nearly constant throughput and delay performance even if the number of active mobile stations in the radio cell rises to 40. Forty users instead of 12 in the pure best effort case can be served with a throughput performance for Interactive applications of 10 kbit/s, while the performance for Background users remains acceptable even in relatively high load situations. These results show that QoS functions in GPRS networks are increasing the application-specific performance significantly and are realizing the capability to serve subscribers and applications with respect to their QoS requirements.

Furthermore the capability of DiffServ to interwork with GPRS QoS functions was examined. To achieve this an IP core network model based on the DiffServ architecture, which is composed of two boundary routers and a cascade of three interior routers, was developed and integrated into the GPRS simulation tool GPRSim. With this model it has been shown that DiffServ is able to support service differentiation, which is done in the radio network based on GPRS/UMTS QoS classes, and also in the core network. There is no difference in the performance of a scenario where the core network influence is fully neglected. If the core network is congested, a significant advantage compared to a best-effort service in the core network is achieved. While a congested core network without IP QoS functions would lead to poor performance for GPRS traffic, DiffServ is able to serve prioritized GPRS traffic with similar performance to that obtained with no influence from the core network. As a result DiffServ is able to interwork with GPRS QoS functions, so that the core network can be seen as transparent and hardly influencing GPRS traffic performance even when the network is congested.

Acronyms

2G	*Second-Generation*
3G	*Third-Generation*
3GPP	*3rd Generation Partnership Project*
3GPP2	*3rd Generation Partnership Project 2*
3G-SGSN	*3rd Generation Serving GPRS Support Node*
8-PSK	*8-Phase-Shift-Keying*
AAA	*Authentication, Authorization and Accounting*
AAC	*Advanced Audio Coding*
ABM	*Asynchronous Balanced Mode*
ABQP	*Aggregate BSS QoS Profile*
AC	*Admission Control*
ACCH	*Associated Control Channel*
ACELP	*Algebraic Codebook Excited Linear Predictive*
ACK	*Acknowledgement*
ADM	*Asynchronous Disconnected Mode*
AGCH	*Access Grant Channel*
ARIB	*Association of Radio Industries and Businesses*
ARQ	*Automatic Repeat Request*
ATM	*Asynchronous Transfer Mode*
AuC	*Authentication Center*
BCCH	*Broadcast Control Channel*
BCS	*Block Check Sequence*
BEC	*Backward Error Correction*
BER	*Bit Error Ratio*
BICC	*Bearer Independent Call Control*
BICN	*Bearer Independent Circuit Network*
BLEP	*Block Error Probability*
BLER	*Block Error Ratio*
BPSCH	*Basic Physical Subchannel*
BS	*Base Station*
BSC	*Base Station Controller*
BSIC	*Base Station Identification Code*
BSN	*Block Sequence Number*
BSS	*Base Station Subsystem*
BSSAP	*Base Station Subsystem Application Part*
BSSGP	*Base Station Subsystem GPRS Protocol*
BTS	*Base Transceiver Station*
BVC	*BSSGP Virtual Connection*
BVCI	*BSSGP Virtual Connection Identifier*
CAC	*Connection Admission Control*

CAMEL	*Customized Applications for Mobile network Enhanced Logic*
CBR	*Constant Bit Rate*
CCCH	*Common Control Channel*
CCS7	*Common Channel Signaling System No. 7*
CDMA	*Code Division Multiple Access*
CEPT	*Conférence Européenne des Administrations des Postes et des Télécommunications*
CGI	*Cell Global Identity*
C/I	*Carrier-to-Interference*
CI	*Cell Identity*
CIF	*Common Intermediate Format*
CLNS	*Connectionless Network Service*
CN	*Core Network*
CNCL	*Communication Networks Class Library*
CNG	*Comfort Noise Generator*
CONS	*Connection Oriented Network Service*
CRC	*Cyclic Redundancy Check*
CRTP	*Compressed RTP*
CS	*Circuit-Switched*
CS	*Coding Scheme*
CSCF	*Call State Control Function*
CSD	*Circuit Switched Data*
CV	*Countdown Value*
CWTS	*Chinese Wireless Telecommunications Standards*
DBPSCH	*Dedicated Basic Physical Subchannel*
DC	*Dedicated Control*
DCCH	*Dedicated Control Channel*
DCS	*Digital Cellular System*
DCT	*Discrete Cosine Transform*
DECT	*Digital Enhanced Cordless Telecommunications*
DHCP	*Dynamic Host Configuration Protocol*
DLC	*Data Link Control*
DLCI	*Data Link Connection Identifier*
DPCM	*Differential Pulse-Code Modulation*
DRX	*Discontinuous Reception*
DSCP	*DiffServ Code Point*
DTP	*Data Transfer Process*
DTX	*Discontinuous Transmission*
ECSD	*Enhanced Circuit-Switched Data*
EDD	*Earliest Due Date*
EDGE	*Enhanced Data rates for GSM Evolution*
EGPRS	*Enhanced General Packet Radio Service*
EI	*Equipment Identity*
EIR	*Equipment Identity Register*
ETSI	*European Telecommunications Standards Institute*
FACCH	*Fast Associated dedicated Control Channel*
FBI	*Final Block Indicator*

FCCH	*Frequency Correction Channel*
FCFS	*First Come First Served*
FCS	*Frame Check Sequence*
FDD	*Frequency Division Duplex*
FDM	*Frequency Division Multiplex*
FDMA	*Frequency Division Multiple Access*
FEC	*Forward Error Correction*
FPB	*First Partial Bitmap*
FR	*Frame Relay*
FSM	*Finite State Machine*
FTP	*File Transfer Protocol*
FUNET	*Finnish University and Research Network*
GC	*General Control*
GERAN	*GSM/EDGE Radio Access Network*
GGSN	*Gateway GPRS Support Node*
GIST	*Graphical Interactive Simulation result Tool*
GMM	*GPRS Mobility Management*
GMM	*Global Multimedia Mobility*
GMSC	*Gateway Mobile-services Switching Center*
GMSK	*Gaussian Minimum Shift Keying*
GOB	*Group Of Blocks*
GPRS	*General Packet Radio Service*
GR	*GPRS Register*
GRX	*GPRS Roaming Exchange*
GSM	*Global System for Mobile communication*
GSM	*Groupe Spéciale Mobile*
GSN	*GPRS Support Node*
GTP	*GPRS Tunneling Protocol*
GTP-C	*GPRS Tunneling Protocol (Control plane)*
GTP-U	*GPRS Tunneling Protocol (User Plane)*
GUI	*Graphical User Interface*
HDLC	*High Level Data Link Control*
HLR	*Home Location Register*
HSCSD	*High-Speed Circuit-Switched Data*
HSS	*Home Subscriber Server*
HTML	*Hypertext Markup Language*
HTTP	*Hypertext Transfer Protocol*
HVS	*Human Visual System*
IDU	*Interface Data Unit*
IE	*Information Element*
IEI	*Information Element Identifier*
IETF	*Internet Engineering Task Force*
IMAP	*Internet Message Access Protocol*
IMS	*IP Multimedia Subsystem*
IMSI	*International Mobile Subscriber Identity*
IMT-2000	*International Mobile Telecommunications at 2000 MHz*
IP	*Internet Protocol*

IPv4	*IP Version 4*
IPv6	*IP Version 6*
IR	*Incremental Redundancy*
ISDN	*Integrated Services Digital Network*
ISO	*International Standards Organization*
ISP	*Internet Service Provider*
ITU	*International Telecommunication Union*
ITU-T	*Telecommunication Standardization Sector of ITU*
IWF	*Interworking Function*
LA	*Location Area*
LA	*Link Adaptation*
LAC	*Location Area Code*
LAI	*Location Area Identity*
LAN	*Local Area Network*
LAPD	*Link Access Procedure on the D-channel*
LDT	*Large Data Transfer*
LLC	*Logical Link Control*
LM	*Location Management*
LPC	*Linear Predictive Coding*
LQC	*Link Quality Control*
LSP	*Linear Spectral Pair*
MAC	*Medium Access Control*
MAP	*Mobile Application Part*
MB	*Macro Block*
MCC	*Mobile Country Code*
MCS	*Modulation and Coding Scheme*
MGCF	*Media Gateway Control Function*
MGCP	*Media Gateway Control Protocol*
MGW	*Media Gateway*
MIME	*Multipurpose Internet Mail Extensions*
MM	*Mobility Management*
MMS	*Multimedia Messaging Service*
MNC	*Mobile Network Code*
MO-C	*Mobile Originated Call*
MoU	*Memorandum of Understanding*
MP3	*MPEG Audio Layer-3*
MPDCH	*Master Packet Data Channel*
MPEG	*Moving Pictures Expert Group*
MPLS	*Multi Protocol Label Switching*
MP-MLQ	*Multi-Pulse Maximum Likelihood Quantization*
MS	*Mobile Station*
MSC	*Mobile-services Switching Center*
MSISDN	*Mobile Station international ISDN number*
MSRN	*Mobile Station Roaming Number*
MT-C	*Mobile Terminated Call*
MTU	*Maximum Transfer Unit*
NACK	*Negative Acknowledgement*
N-CDMA	*Narrowband-CDMA*
NGN	*Next Generation Network*

NPB *Next Partial Bitmap*
NRT *Non Real-Time*
N-SAPI *Network layer Service Access Point Identifier*
NSS *Network and Switching Subsystem*
Nt *Notification*
NTSC *National Television System Committee*
OMC *Operating and Maintenance Center*
OSI *Open Systems Interconnection*
OSS *Operation Subsystem*
PACCH *Packet Associated Control Channel*
PAD *Packet Assembly Disassembly*
PAGCH *Packet Access Grant Channel*
PAL *Phase-Alternating Line*
PBCCH *Packet Broadcast Control Channel*
PBSN *Partial Bitmap Sequence Number*
PC *Personal Computer*
PCCCH *Packet Common Control Channel*
PCH *Paging Channel*
PCM *Pulse Code Modulation*
PDA *Personal Digital Assistant*
PDC *Personal Digital Cellular*
PDCH *Packet Data Channel*
PDN *Packet Data Network*
PDP *Packet Data Protocol*
PDTCH *Packet Data Traffic Channel*
PDU *Protocol Data Unit*
PFC *Packet Flow Context*
PFI *Packet Flow Identifier*
PHB *Per-Hop Behavior*
PI *Protocol Interpreter*
PILC *Performance Implications of Link Characteristics*
PL *Physical Layer*
PLL *Physical Link Layer*
PLMN *Public Land Mobile Network*
PNCH *Packet Notification Channel*
POP3 *Post Office Protocol version 3*
PPCH *Packet Paging Channel*
PRACH *Packet Random Access Channel*
PS *Packet-Switched*
PSDN *Public Switched Data Network*
PSTN *Public Switched Telephony Network*
PTCH *Packet Traffic Channel*
PTM *Point-to-Multipoint*
PTM-G *PTM Group Call*
PTM-M *PTM Multicast*
P-TMSI *Packet Temporary Mobile Subscriber Identity*
PTP *Point-To-Point*
PTT *Postal, Telephone and Telegraph administration*
Q10 *Quantization Level 10*

QCIF	*Quarter Common Intermediate Format*
QoS	*Quality of Service*
RA	*Routing Area*
RAB	*Radio Access Bearer*
RAC	*Routing Area Code*
RACH	*Random Access Channel*
RADIUS	*Remote Authentication Dial In User Service*
RAI	*Routing Area Identity*
RAN	*Radio Access Network*
RANAP	*Radio Access Network Application Part*
RB	*Reported Bitmap*
RBB	*Received Block Bitmap*
RED	*Random Early Detection*
RF	*Radio Frequency*
RLC	*Radio Link Control*
RNC	*Radio Network Controller*
ROHC	*Robust Header Compression*
RR	*Round Robin*
RRBP	*Relative Reserved Block Period*
RRC	*Radio Resource Control*
RSS	*Radio Subsystem*
RT	*Real-Time*
RTCP	*Real-Time Control Protocol*
RTP	*Real-time Transmission Protocol*
RTSP	*Real-Time Streaming Protocol*
SABM	*Set Asynchronous Balance Mode*
SACCH	*Slow Associated dedicated Control Channel*
SACK	*Selective Acknowledgement*
SAP	*Service Access Point*
SAPI	*Service Access Point Identifier*
SB	*Synchronization downlink Burst*
SBPSCH	*Shared Basic Physical Subchannel*
SCH	*Synchronization Channel*
SDCCH	*Stand-alone Dedicated Control Channel*
SDL	*Specification and Description Language*
SDU	*Service Data Unit*
SGSN	*Serving GPRS Support Node*
SI	*Stall Indicator*
SID	*Silence Insertion Descriptor*
SIM	*Subscriber Identity Module*
SIP	*Session Initiation Protocol*
SLA	*Service Level Agreement*
SM	*Session Management*
SMG	*Special Mobile Group*
SMIL	*Synchronised Multimedia Integration Language*
SMS	*Short Message Service*
SMTP	*Simple Mail Transfer Protocol*
SNDCP	*Sub-Network Dependent Convergence Protocol*
SNS	*Sequence Number Space*

SPEETCL	*SDL Performance Evaluation Tool Class Library*
SSA	*Services and System Aspects*
SSL	*Secure Socket Layer*
SSN	*Starting Sequence Number*
TA	*Terminal Aspects*
TA	*Timing Advance*
TBF	*Temporary Block Flow*
TCH	*Traffic Channel*
TCP	*Transport Control Protocol*
TD-SCDMA	*Time Division—Synchronous Code Division Multiple Access*
TDD	*Time Division Duplex*
TDM	*Time Division Multiplex*
TDMA	*Time Division Multiple Access*
TFI	*Temporary Flow Identity*
TFT	*Traffic Flow Template*
TI	*Transaction Identifier*
TIA	*Telecommunication Industry Association*
TID	*Tunnel Identifier*
TLLI	*Temporary Logical Link Identifier*
TLS	*Transport Layer Security*
TMN	*Telecommunications Management Network*
TMSI	*Temporary Mobile Subscriber Identity*
TN	*Timeslot Number*
TOS	*Type of Service*
TRX	*Transceiver*
TSG	*Technical Specification Group*
TTA	*Telecommunication Technology Association*
TTC	*Telecommunication Technology Committee*
UDP	*User Datagram Protocol*
UMTS	*Universal Mobile Telecommunication System*
URL	*Unified Resource Locator*
USF	*Uplink State Flag*
USSD	*Unstructured Supplementary Service Data*
UTRAN	*UMTS Terrestrial Radio Access Network*
UWC	*Universal Wireless Communications*
VAD	*Voice Activity Detection*
VBR	*Variable Bit Rate*
VCC	*Visited Country Code*
VLR	*Visitor Location Register*
VMSC	*Visited MSC*
VNDC	*Visited NDC*
VoIP	*Voice over IP*
VPN	*Virtual Private Network*
VQEG	*Video Quality Expert Group*
VSN	*Visited Subscriber Number*
WAE	*Wireless Application Environment*
WAN	*Wide Area Network*
WAP	*Wireless Application Protocol*
WBXML	*WAP Binary Extensible Markup Language*

W-CDMA *Wideband Code Division Multiple Access*

WDP *Wireless Datagram Protocol*

WFQ *Weighted Fair Queueing*

WML *Wireless Markup Language*

WP-TCP *Wireless Profiled TCP*

WRR *Weighted Round Robin*

WS *Window Size*

WSP *Wireless Session Protocol*

WTA *Wireless Telephony Application*

WTLS *Wireless Transport Layer Security*

WTP *Wireless Transaction Protocol*

WWW *World Wide Web*

XML *Extensible Markup Language*

Bibliography

[1] 3GPP TCG SSA. *Security-related Network Functions (3GPP TS 43.020 version 4.0.0 Release 4)*. Technical Specification, 3rd Generation Partnership Project, November 2001.

[2] 3GPP TSG CN. *Handover Procedures (3GPP TS 23.009 version 5.0.0 Release 5)*. Technical Specification, 3rd Generation Partnership Project, December 2001.

[3] 3GPP TSG CN. *Mobile radio interface signalling layer 3; General aspects (3GPP TS 24.007 version 4.1.0 Release 4)*. Technical Specification, 3rd Generation Partnership Project, December 2001.

[4] 3GPP TSG CN. *Numbering, addressing and identification (3GPP TS 23.003 version 5.2.0 Release 5)*. Technical Specification, 3rd Generation Partnership Project, December 2001.

[5] 3GPP TSG CN. *Bearer-independent Circuit-switched Core Network; Stage 2 (3GPP TS 23.205 version 5.1.0 Release 5)*. Technical Specification, 3rd Generation Partnership Project, March 2002.

[6] 3GPP TSG CN. *Customized Applications for Mobile Network Enhanced Logic (CAMEL) Phase 3 - Stage 2 (3G TS 23.078 version 4.4.0 Release 4)*. Technical Specification, 3rd Generation Partnership Project, March 2002.

[7] 3GPP TSG CN. *Digital cellular telecommunications system (Phase 2+); General Packet Radio Service (GPRS); Mobile Station (MS) – Serving GPRS Support Node (SGSN); Logical Link Control (LLC) layer specification (3GPP TS 44.064 version 5.1.0 Release 5)*. Technical Specification, 3rd Generation Partnership Project, March 2002.

[8] 3GPP TSG CN. *Digital cellular telecommunications system (Phase 2+); General Packet Radio Service (GPRS); Mobile Station (MS) – Serving GPRS Support Node (SGSN); Subnetwork Dependent Convergence Protocol (SNDCP) (3GPP TS 44.065 version 5.0.0 Release 5)*. Technical Specification, 3rd Generation Partnership Project, June 2002.

[9] 3GPP TSG CN. *General Packet Radio Service (GPRS); GPRS Tunnelling Protocol (GTP) across the Gn and Gp Interface (3GPP TS 29.060 version 5.0.1 Release 5)*. Technical Specification, 3rd Generation Partnership Project, January 2002.

[10] 3GPP TSG CN. *General Packet Radio Service (GPRS); Interworking between the Public Land Mobile Network (PLMN) supporting GPRS and Packet Data Networks (3GPP TS 29.061 version 5.1.0 Release 5)*. Technical Specification, 3rd Generation Partnership Project, January 2002.

[11] 3GPP TSG CN. *Mobile radio interface layer 3 specification; Core Network Protocols (3GPP TS 24.008 version 5.2.0 Release 5)*. Technical Specification, 3rd Generation Partnership Project, June 2002.

[12] 3GPP TSG CN. *Unstructured Supplementary Service Data (3G TS 24.090 version 4.0.0 Release 4)*. Technical Specification, 3rd Generation Partnership Project, March 2002.

[13] 3GPP TSG GERAN. *Channel Coding (3GPP TS 45.003 version 5.6.0 Release 5)*. Technical Specification, 3rd Generation Partnership Project, June 2002.

[14] 3GPP TSG GERAN. *Digital cellular telecommunications system (Phase 2+); Physical layer on the radio path; General description (3GPP TS 45.001 version 5.4.0 Release 5)*. Technical Specification, 3rd Generation Partnership Project, June 2002.

[15] 3GPP TSG GERAN. *General Packet Radio Service (GPRS); Base Station System (BSS) – Serving GPRS Support Node (SGSN); BSS GPRS Protocol (BSSGP) (3GPP TS 48.018 version 5.4.0 Release 5)*. Technical Specification, 3rd Generation Partnership Project, July 2002.

[16] 3GPP TSG GERAN. *General Packet Radio Service (GPRS); Mobile Station (MS) – Base Station System (BSS) interface; Radio Link Control / Medium Access Control (RLC/MAC) protocol (3GPP TS 44.060 version 5.2.0 Release 5)*. Technical Specification, 3rd Generation Partnership Project, July 2002.

[17] 3GPP TSG GERAN. *General Packet Radio Service (GPRS); Overall description of the GPRS radio interface; Stage 2 (3GPP TS 43.064 version 5.0.0 Release 5)*. Technical Specification, 3rd Generation Partnership Project, April 2002.

[18] 3GPP TSG GERAN. *Overall Description - Stage2 (3GPP TS 43.051 version 5.6.0 Release 5)*. Technical Specification, 3rd Generation Partnership Project, April 2002.

[19] 3GPP TSG GERAN. *Radio subsystem link control (3GPP TS 45.008 version 5.4.0 Release 5)*. Technical Specification, 3rd Generation Partnership Project, June 2002.

[20] 3GPP TSG GERAN. *Radio Transmission and Reception (3GPP TS 45.005 version 5.4.0 Release 5)*. Technical Specification, 3rd Generation Partnership Project, June 2002.

[21] 3GPP TSG RAN. *Radio Resource Control (RRC) (3GPP TS 25.331 version 5.0.0 Release 5)*. Technical Specification, 3rd Generation Partnership Project, March 2002.

[22] 3GPP TSG SSA. *Architecture for an All IP Network (3GPP TR 23.922 version 1.0.0)*. Technical Report, 3rd Generation Partnership Project, December 1999.

[23] 3GPP TSG SSA. *Circuit Teleservices supported by a Public Land Mobile Network (PLMN) (3GPP TS 22.003 version 5.0.0 Release 5)*. Technical Specification, 3rd Generation Partnership Project, December 2001.

[24] 3GPP TSG SSA. *Digital cellular telecommunications system (Phase 2+) (GSM); Universal Mobile Telecommunications System (UMTS); Principles of circuit telecommunication services supported by a Public Land Mobile Network (PLMN) (3G TS 22.001 version 4.2.0 Release 4).* Technical Specification, 3rd Generation Partnership Project, July 2001.

[25] 3GPP TSG SSA. *General Packet Radio Service (GPRS); Service description, Stage 1 (3G TS 22.060 version 5.0.0 Release 5).* Technical Specification, 3rd Generation Partnership Project, October 2001.

[26] 3GPP TSG SSA. *General UMTS Architecture (3GPP TS 23.101 version 4.0.0 Release 4).* Technical Specification, 3rd Generation Partnership Project, April 2001.

[27] 3GPP TSG SSA. *Services and Service Capabilities (3G TS 22.105 version 5.0.0 Release 5).* Technical Specification, 3rd Generation Partnership Project, October 2001.

[28] 3GPP TSG SSA. *General on Supplementary Services (3GPP TS 22.004 version 4.1.0 Release 4).* Technical Specification, 3rd Generation Partnership Project, March 2002.

[29] 3GPP TSG SSA. *General Packet Radio Service (GPRS); Service description; Stage 2 (3GPP TS 23.060 version 5.0.0 Release 5).* Technical specification, 3rd Generation Partnership Project, January 2002.

[30] 3GPP TSG SSA. *QoS Concept and Architecture (3GPP TS 23.107 version 5.3.0 Release 5).* Technical Specification, 3rd Generation Partnership Project, January 2002.

[31] Aixcom GmbH. `http://www.aixcom.com/Produkte_e.php`.

[32] M. Arango, A. Dugan, I. Elliott, C. Huitema, S. Pickett. *Media Gateway Control Protocol, Request for Comments 2705.* Technical Report, Internet Engineering Task Force (IETF), October 1999.

[33] M. Arlitt, C. Williamson. *A Synthetic Workload Model for Internet Mosaic Traffic.* In *Proc. of the 1995 Summer Computer Simulation Conference*, pp. 24–26, Ottawa, Canada, 1995.

[34] U. Black. *QoS in Wide Area Networks.* Prentice Hall, Upper Saddle River, N.J., 2000.

[35] S. Blake, D. Black, M. Carlson, E. Davis, Z. Wang, W. Weiss. *An Architecture for Differentiated Services, Request for Comments 2475.* Technical Report, Internet Engineering Task Force (IETF), December 1998.

[36] K. J. Blyth, A. R. Cook. *Designing a GPRS Roaming Exchange Service.* In *Proc. of the Second International Conference on 3G Mobile Communication Technologies*, London, UK, March 2001.

[37] C. Bormann, C. Burmeiste, M. Degermark, H. Fukushima, H. Hannu, L-E. Jonsson, R. Hakenberg, T. Koren, K. Le, Z. Liu, A. Martensson, A. Miyazak, K. Svanbro, T. Wiebke, T. Yoshimura, H. Zheng. *Robust Header Compression: Framework and four Profiles: RTP, UDP, ESP, and uncompressed, Request for Comments 3095.* Technical Report, Internet Engineering Task Force (IETF), July 2001.

[38] C. Bormann, L. Cline, G. Deisher, T. Gardos, C. Maciocco, D. Newell, J. Ott, G. Sullivan, S. Wenger, C. Zhu. *RTP Payload Format for the 1998 Version of ITU-T Rec. H.263 Video (H.263+), Request for Comments 2429.* Technical Report, Internet Engineering Task Force (IETF), October 1998.

[39] B. Braden, L. Zhang, S. Berson, S. Herzog, S. Jamin. *Resource Reservation Protocol, Request for Comments 2205.* Technical Report, Internet Engineering Task Force (IETF), September 1997.

[40] R. Braden, D. Clark, S. Shenker. *Integrated Services in the Internet Architecture, Request for Comments 1633.* Technical Report, Internet Engineering Task Force (IETF), June 1994.

[41] Paul T. Brady. *A Model for Generating On-Off Speech Patterns in Two-Way Conversation.* The Bell System Technical Journal, No. 9, Vol. 48, pp. 2445–2472, September 1969.

[42] G. Brasche, B. Walke. *Concepts, Services and Protocols of the New GSM Phase 2+ General Packet Radio Service.* IEEE Communications Magazine, Vol. 35, 1997.

[43] H. K. Choi, J.O. Limb. *A Behavioral Model of Web Traffic.* In *Proc. of the 7th International Conference on Network Protocols (ICNP 99)*, Ontario, Canada, October 1999.

[44] I. Costa, L. Dell'Uomo. *Quality of Service in Future 4G Mobile Wireless Systems.* In *Proc. of the IEEE International Conference on Third Generation Wireless and Beyond (3Gwireless '01)*, pp. 502–507, June 2001.

[45] M. Crispin. *Internet Message Access Protocol – Version 4, Request for Comments 2060.* Technical Report, Internet Engineering Task Force (IETF), December 1996.

[46] P. Decker. *A Packet Radio Protocol Proposed for the GSM Mobile Radio Network.* In *Workshop on Mobile Multimedia Communication (MoMuC-1)*, Tokyo, IEEE, December 1993.

[47] P. Decker. *Entwurf und Leistungsbewertung hybrider Fehlersicherungsprotokolle für paketorientierte Sprach- und Datendienste in GSM-Mobilfunksystemen.* Dissertation, RWTH Aachen, Lehrstuhl für Kommunikationsnetze, 1995.

[48] Shuang Deng. *Traffic Characteristics of Packet Voice.* IEEE International Conference on Communications, pp. 1369–1374, June 1995.

[49] A. Dutta-Roy. *The Cost of Quality in Internet-style Networks.* IEEE Spectrum, No. 0018-9235/00, pp. 57–62, September 2000.

[50] J. Eberspächer, H.-J. Vögel, C. Bettstetter. *GSM Switching, Services and Protocols, 2nd Edition.* John Wiley & Sons, Chichester, 2001.

[51] E. O. Elliott. *Estimates of error rates for codes on burst-noise channels.* The Bell System Technical Journal, Vol. 42, pp. 1977–1997, September 1963.

[52] I. Elsen, F. Hartung, U. Horn, M. Kampmann, L. Peters. *Streaming Technology in 3G Mobile Communication Systems.* IEEE Computer, Vol. 34, No. 9, pp. 46–52, 2001.

[53] ETSI. *Overall Requirements on the Radio Interface of the Universal Mobile Telecommunication System (UMTS) (ETSI TR 04-01/SMG-050401).* Technical Report, European Telecommunications Standards Institute, September 1996.

[54] ETSI 3GPP. *Selection Procedures for the Choice of Radio Transmission Technologies of the Universal Mobile Telecommunication System UMTS (3G TR 101 112).* Technical Report ETSI TR 101 112, European Telecommunications Standards Institute, April 1998.

[55] ETSI SMG 2. *TDoc 1549/99, Concept Proposal for EGPRS-136, Revision 1.6.* Technical report, European Telecommunications Standards Institute.

[56] R. Fielding, J. Gettys, J. Mogul, H. Frystyk, T. Berners-Lee. *Hypertext Transfer Protocol – HTTP/1.1, Request for Comments 2068.* Technical Report, Internet Engineering Task Force (IETF), January 1997.

[57] U. Fornelfeld, P. Stuckmann, Adolfo Bayo, Rosalino Critelli. *Analysis of the Influence of Packet-data and Voice Traffic on the Interference Situation in Multicellular GSM/(E)GPRS Networks.* In *Proc. of 10th Aachen Symposium on Signal Theory*, pp. 377–382, Aachen, Germany, VDE Verlag, September 2001.

[58] WAP Forum. *Wireless Profiled TCP.* Technical Report WAP-225-TCP-200103331-a, WAP Forum, March 2001.

[59] A. Furuskär, S. Mazur, F. Müller, H. Olofsson. *EDGE: Enhanced Data Rates for GSM and TDMA/136 Evolution.* IEEE Personal Communications, pp. 56–65, June 1999.

[60] A. Furuskär, H. Olofsson. *Aspects of Introducing EDGE in Existing GSM Networks.* Number 1, pp. 28–37, Stockholm, Sweden, Ericsson Radio Systems AB, 1999.

[61] E. N. Gilbert. *Capacity of a burst-noise channel.* Bell System Technical Journal, Vol. 39, pp. 1253–1266, September 1960.

[62] H. Gudding. *Capacity Analysis of GPRS.* Master thesis, NTNU Norwegian University of Science and Technology, Trondheim, Norway, March 2000.

[63] R. Guerin, H. Ahmadi, M. Naghshineh. *Equivalent Capacity and its Application to Bandwidth Allocation in High-Speed Networks.* IEEE Journal on Selected Areas on Communications, 9(7), pp. 868–881, 1991.

[64] T. Halonen, J. Romero, J. Melero. *GSM, GPRS and EDGE Performance — Evolution Towards 3G/UMTS.* John Wiley & Sons, Chichester, 2002.

[65] M. Handley, H. Schulzrinne, E. Schooler, J. Rosenberg. *SIP: Session Initiation Protocol, Request for Comments 2543.* Technical Report, Internet Engineering Task Force (IETF), March 1999.

[66] H. Heffes, D.M. Lucantoni. *A Markov Modulated Characterization of Packetized Voice and Data Traffic and Related Statistical Multiplexer Performance.* IEEE Journal on Selected Areas on Communications, 4(6), pp. 856–867, 1986.

[67] Christoph Herrmann. *Stochastic Models for ATM Concepts (in German).* Aachener Beiträge zur Mobil- und Telekommunikation. Verlag der Augustinus Buchhandlung, Aachen, 1995. Dissertation.

[68] F. Hillebrandt. *The Creation of Global Mobile Communication.* John Wiley & Sons, Chichester, 2002.

[69] ISO/IEC JTC1/SC29 WG11. *MPEG-2 Advanced Audio Coding.* International Standard IS 13818-7, International Organization for Standardization, Geneva, Switzerland, 1997.

[70] ITU-T. *Data Compression Procedures for Data Circuit Terminating Equipment (DCE) Using Error Correction Procedures.* ITU-T Recommendation V.42bis, International Telecommunication Union – Telecommunication Standardization Sector, Geneva, Switzerland, 1990.

[71] ITU-T. *Recommendation H.261, Series H: Audiovisual and Multimedia Systems - Video codec for audiovisual services at p x 64 kbit/s.* Technical Report H.261, March 1993.

[72] ITU-T. *Packet-switched Signalling System between Public Networks and Providing Data Transmission Services.* ITU-T Recommendation X.75, International Telecommunication Union – Telecommunication Standardization Sector, Geneva, Switzerland, October 1996.

[73] ITU-T. *Speech coders: Dual Rate Speech Coder for Multimedia Communications transmitting at 5.3 and 6.3 kbit/s.* Technical Report G.723, March 1996.

[74] ITU-T. *Draft Text of Recommendation H.263 Version 2 (H.263+) for Decision.* Technical Report H.263, September 1997.

[75] ITU-T. *Gateway Control Protocol.* Technical Report H.248, June 2000.

[76] ITU-T. *Packet-based Multimedia Communications System.* Technical Report H.323, November 2000.

[77] ITU-T SG 2. *Terms and Definitions Related to Quality of Service and Network Performance Including Dependability.* ITU-T Recommendation E.800, International Telecommunication Union – Telecommunication Standardization Sector, Geneva, Switzerland, 1994.

[78] V. Jacobson. *Compressing TCP/IP Headers for Low-Speed Serial Links, Request for Comments 1144.* Technical Report, Internet Engineering Task Force (IETF), February 1990.

[79] D. Johnson, C. Perkins. *Mobility Support in IPv6, IETF Internet Draft.* Technical Report, Internet Engineering Task Force (IETF), July 2001.

[80] R. Kalden, I. Meirick, M. Meyer. *Wireless Internet Access Based on GPRS.* IEEE Personal Communications, Vol. 4, 2000.

[81] I. Karatzas, S. Shreve. *Brownian Motion and Stochastic Calculus.* Second Edition, Springer-Verlag New York, 1991.

[82] L. Kleinrock. *Queueing Systems*, Vol. I: *Theory.* John Wiley & Sons, New York/London/Sydney/Toronto, 1975.

[83] R. Koodli, M. Puuskari. *Supporting Packet-Data QoS in Next-Generation Cellular Networks.* IEEE Communications Magazine, Vol. 39, pp. 180–188, February 2001.

[84] D. Lucantoni. *The BMAP/D/1 Queue: A Tutorial.* Models and Techniques for Performance Evaluation of Computer and Communication Systems, L. Donatiello and R. Nelson (Editors), pp. 331–358, 1993.

[85] M. Junius et al. *CNCL: a C++ library for event driven simulation, statistical evaluation and random number generators and distributions.* Technical Report, Communication Networks, Aachen University of Technology, 1993.

[86] Mounir Merhi. *Modeling and Implementation of a Mobility Model for the Simulative Performance Assessment of the General Packet Radio Service (GPRS).* Master's thesis, Aachen University, Chair of Communication Networks, November 1999.

[87] D. Mitton, M. St.Johns, S. Barkley, D. Nelson, B. Patil, M. Stevens, B. Wolff. *Authentication, Authorization, and Accounting: Protocol Evaluation, Request for Comments 3127.* Technical Report, Internet Engineering Task Force (IETF), June 2001.

[88] M. Mouly, M.-B. Pautet. *The GSM System for Mobile Communications, ISBN 2-9507190-0-7.* published by the authors, 49, rue Louise Bruneau, F-91129 Palaiseau, France, 1992.

[89] J. Myers, M. Rose. *Post Office Protocol — Version 3, Request for Comments 1939.* Technical Report, Internet Engineering Task Force (IETF), May 1996.

[90] K. Nichols, S. Blake, F. Baker, D. Black. *Definition of the Differentiated Services Field in the IPv4 and IPv6 Headers, Request for Comments 2474.* Technical Report, Internet Engineering Task Force (IETF), December 1998.

[91] W. Olzem. *Integration of protocols specified in SDL into an event-driven simulation environment.* Technical report, RWTH Aachen, Lehrstuhl für Kommunikationsnetze, Kopernikusstr. 16, D-52074 Aachen, Germany, January 1995.

[92] V. Paxson. *Empirically-Derived Analytic Models of Wide-Area TCP Connections.* IEEE/ACM Transactions on Networking, Vol. 2, No. 4, pp. 316–336, August 1994.

[93] V. Paxson, S. Floyd. *Wide-Area Traffic: The Failure of Poisson Modeling*, June 1995.

[94] C. Perkins. *IP Mobility Support, Request for Comments 2002.* Technical Report, Internet Engineering Task Force (IETF), October 1996.

[95] J. Postel. *Transmission Control Protocol, Request for Comments 793.* Technical Report, Internet Engineering Task Force (IETF), September 1981.

[96] J. Postel. *Simple Mail Transfer Protocol, Request for Comments 821.* Technical Report, Internet Engineering Task Force (IETF), August 1982.

[97] J. Postel, J. Reynolds. *File Transfer Protocol, Request for Comments 959.* Technical Report, Internet Engineering Task Force (IETF), Oktober 1985.

[98] O. Rose, M. Frater. *A comparison of models for VBR video traffic sources in B-ISDN.* IFIP transactions C-24: Broadband communications II, pp. 275–287, 1994.

[99] E. Rosen, A. Viswanathan, R. Callon. *Multiprotocol Label Switching Architecture, Request for Comments 3031.* Technical Report, Internet Engineering Task Force (IETF), January 2001.

[100] H. Schulzrinne, S. Casner, R. Frederick, V. Jacobson. *RTP: A Transport Protocol for Real-Time Applications, Request for Comments 1889.* Technical Report, Internet Engineering Task Force (IETF), March 2001.

[101] H. Schulzrinne, A. Rao, R. Lanphier. *Real Time Streaming Protocol (RTSP), Request for Comments 2326.* Technical Report, Internet Engineering Task Force (IETF), April 1998.

[102] N. Sollenberger, N. Seshadri, R. Cox. *The Evolution of IS-136 TDMA for Third-Generation Wireless Services.* IEEE Personal Communications, Vol. 6, 1999.

[103] M. Steppler. *Performance Analysis of Communication Systems Formally Specified in SDL.* In *Proc. of the First International Workshop on Simulation and Performance '98 (WOSP '98)*, pp. 49–62, 1998.

[104] P. Stuckmann. *Quality of Service Management in GPRS-based Radio Access Networks.* Telecommunication Systems 19:3, Kluwer Academic Publishers, pp. 515–546, March 2002.

[105] P. Stuckmann, H. Finck, T. Bahls. *A WAP Traffic Model and its Appliance for the Performance Analysis of WAP over GPRS.* In *Proc. of the IEEE International Conference on Third Generation Wireless and Beyond (3Gwireless '01)*, San Francisco, USA, June 2001.

[106] P. Stuckmann, F. Müller. *GPRS Radio Network Capacity Considering Coexisting Circuit-Switched Traffic Sources.* In *Proc. of European Conference on Wireless Technology (ECWT 2000)*, 2000.

[107] P. Stuckmann, F. Müller. *Quality of Service Management in GPRS Networks.* In *Proc. of the IEEE International Conference on Networking (ICN '01)*, July 2001.

[108] M. Taferner, E. Bonek. *Wireless Internet Access over GSM and UMTS.* Springer, Berlin, 2002.

[109] J. Tisal. *The GSM Network.* 2nd Edition, John Wiley & Sons, Chichester, 2001.

[110] W3C. *Synchronized Multimedia Integration Language (SMIL) Boston Specification.* Working Draft SMIL 2.0, September 2001.

[111] B. Walke. *Mobile Radio Networks — Networking, Protocols and Traffic Performance.* 2nd Edition, John Wiley & Sons, Chichester, 2001.

[112] B. Walke, W. Mende, P. Decker, R. Crumbach. *The Performance of CELLPAC: A Packet Radio Protocol Proposed for the GSM Mobile Radio Network.* In *Mobile Radio Conference*, pp. 57–63, Nice, France, November 1991.

[113] B. Walke, W. Mende, G. Hatziliadis. *CELLPAC, a packet radio protocol applied to the cellular GSM mobile radio network.* In *Proc. of the 41st IEEE Vehicular Technology Conference*, pp. 408–413, St. Louis, Missouri, USA, May 1991.

[114] J. Wigard, P. Mogensen. *A Simple Mapping from C/I to FER and BER for a GSM Type of Air-Interface.* In *Proc. of the International Symposium on Personal, Indoor and Mobile Radio Communications (PIMRC)*, pp. 78–82, 1996.

[115] J. Wigard, T. T. Nielsen, P. H. Michaelsen, P. Mogensen. *BER and FER Prediction of Control and Traffic Channels for a GSM Type of Air-Interface.* In *Proc. of the Vehicular Technology Conference (VTC)*, pp. 1588–1592, 1998.

[116] Sherman Willinger, Taqqu. *Self-similarity through high-variability: Statistical analysis of ethernet LAN traffic at the source level.* IEEE/ACM Transactions on Networking, Vol. 5, pp. 71–86, Feb 1997.

[117] Wireless Application Protocol Forum. *Wireless Application Protocol - Wireless Application Environment Overview.* Technical report, 1999.

[118] Wireless Application Protocol Forum. *Wireless Application Protocol Architecture Specification.* Technical report, 1999.

[119] Wireless Application Protocol Forum. *Wireless Application Protocol - Multimedia Messaging Service Architecture Overview Specification, Version 25.* Technical report, 2001.

[120] Wireless Application Protocol Forum. *Wireless Application Protocol - WAP 2.0 Technical White Paper.* Technical report, 2001.

[121] Wireless Application Protocol Forum. *Wireless Application Protocol 2.0 - Wireless Application Protocol Architecture Specification.* Technical report, 2001.

[122] Y. Yatsuzuka. *Highly Sensitive Speech Detector and High-Speed Voiceband Discrimination in SDI-ADPCM system.* IEEE Transactions on Communications, No. 4, Vol. 30, pp. 739–750, April 1982.

[123] M. Zaddach, K. Heidtmann. *Measurement and Traffic Characterization of H.26x-coded Video Streams.* In *Proc. of the 11th GI/ITG Conference on Measuring, Modelling and Evaluation of Computer and Communication Systems.* VDE-Verlag, Berlin, 2001.

[124] C. Zhu. *RTP Payload Format for H.263 Video Streams, Request for Comments 2190.* Technical Report, Internet Engineering Task Force (IETF), September 1997.

Index

Symbols

A

B

C

F

G